D0838639

Introduction to Flavonoids
The Gordon and Breach Publishing Group

Flavonoids of the Sunflower Family (Asteraceae)
Springer-Vertag

Hawai'i's Native Plants

Dr. Bruce A. Bohm

MUTUAL PUBLISHING

ISBN 1-56647-666-6
Library of Congress Catalog
Card Number 2004103629

First Printing, September 2004
1 2 3 4 5 6 7 8 9

Design by Angela Wu-Ki

All photos by Bruce A. Bohm
unless otherwise noted

Mutual Publishing
1215 Center Street, Suite 210
Honolulu, Hawai'i 96816
Ph (808)732-1709
Fax (808)734-4094
Email: mutual@mutualpublishing.com
www.mutualpublishing.com

Printed in Korea

DEDICATION

This book is dedicated to Sherwin Carlquist

whose many accomplishments include *Hawaii: a Natural History*

and seminal investigations on the origin of the

silversword alliance. These and other works

set the scene and the standard for the following generations of investigators.

TABLE OF CONTENTS

PREFACE

It as been said that a society should be judged on how it treats its weakest, most vulnerable members. If we consider the Earth's centers of biodiversity as among the most vulnerable places on the planet, then as a society we are not doing very well. When considering the broader aspects of endemism, it is sobering to find that a majority of the Earth's native species of vascular plants occupy little more than 10 percent of the total land area, and that less than half of this area is even partially protected. The pressure to provide accommodations, and even barely adequate food, for a human population that appears to be increasing without control, as well as the extraction of non-renewable resources to support this growth, has driven the Earth's biota to the brink of disaster.

Nowhere is the impact of human activities more evident than on oceanic islands—although the ravages of human intervention may not be obvious to the untrained eye. There are extreme cases where over-harvesting of forests followed by serious loss of topsoil through erosion, as seen on Madagascar or St. Helena, is painfully obvious. The problem in the Hawaiian Islands is that large expanses of clearcut wasteland, weed-covered fields, and native forests are not immediately obvious, and the first-time visitor will not notice that significant tracts are covered by invasive alien species. For the most part, the visitor sees a beautiful, lush "tropical paradise" (just read the tourist brochures). I have spoken with many visitors to the islands who were not aware that unless they made a special effort, they may not see a truly Hawaiian plant during their entire stay.

Does anyone care? Many visitors may well not care; they visit the islands to enjoy the climate, the beaches, and the genial atmosphere characteristic of the place. Should anyone care? Since the majority of visitors to the islands come from one of the other United States, it should be a concern to them that part of their country is in serious need of drastic conservation efforts! It is the conservationist's desire, and in fact, obligation, to inform and educate. This is not to say that visitors from other parts of the world should be any less concerned. It is true that the Hawaiian Islands are part of the United States, but they are also part of the only world we have. It is no less true in the Hawaiian Islands than anywhere else on Earth, that if one is not part of the solution, one could well be part of the problem.

In addition to hands-on work by local conservationists, education of the visiting public is also a critical factor in attempting to protect sensitive environments from further harm. In discussing conservation issues on (but certainly not limited to) the Hawaiian Islands, Sherwin Carlquist (1995) wrote in the introductory chapter to *Hawaiian Biogeography. Evolution on a Hot Spot Archipelago* that we should be…"concerned with putting information about endangered species in the hands of the public—simple, appealing facts they can associate with the name of a plant or animal. Information of this sort is essential to public support of conservation efforts, because *the public supports conservation of plants and animals about which it knows something.* Species about which the public knows nothing or about which the public has no visual image are unlikely to be conserved." Easily available sources of information on Hawaiian plants include a variety of books dealing with such subjects as trail-side plants of Hawaii, plants of the seashore, ethnobotany (cultural uses of plants), and trees of Hawaii, all of which are generally available at book and magazine vendors throughout the islands. A list of some of the available literature appears in Chapters One and Two.

It is my hope that the present book on native Hawaiian species will add to the information available to the public. The book's primary intent is to demonstrate that some native species on the Hawaiian Islands can be found with comparatively little effort. In many cases, interesting native species can be found along well-traveled walkways easily reached by automobile—extensive (or expensive) expeditions in the field are not necessary. The more adventuresome visitor will also be richly rewarded by hiking some of the more challenging trails. All that is needed is a note here and there to point in the direction of something interesting: common, uncommon, rare, and perhaps even endangered. That is the reason for writing this book. It is my hope that the background information provided on many of these species will add to the visitors' pleasure and appreciation that what they are seeing in the Hawaiian Islands can be found in Nature nowhere else.

ACKNOWLEDGEMENTS

The following people have made significant contributions to the preparation of illustrations for this book: my long time companion, illustrator, and wife, Lesley Bohm, whose work appears within; Mike Hawkes of the University of British Columbia for photographs of some Hawaiian ferns, *Clermontia parviflora*, and a photograph of observatories on Mauna Kea; Bruce Baldwin of the Jepson Herbarium, University of California, Berkeley for photographs of *Carlquistia muirii* and *Sanicula* species; Anne Westerbergh of Umeå Universitry, Umeå, Sweden for photographs of *Silene*; Ken Marr of the British Columbia Provincial Museum for photographs of *Lysimachia*, island cottons, and several lobelioids; and Fred Ganders of the University of British Columbia for the *Bidens cosmoides* photograph. Thanks also go to *Pacific Science* (University of Hawai'i Press) for permission to reproduce the illustration of *Lysimachia iniki*, which was drawn by my wife. I am also grateful to Charlotte Lindqvist of the Natural History Museums and Botanical Garden at the University of Oslo, Norway for bringing to my attention certain publications and for comments on the Hawaiian mints; and to Jane Villa-Lobos, of the *Plant Talk* staff, for permission to use her tabulated data on hot spots. Thanks to Dr. Gerry Carr of the Department of Botany, University of Hawai'i for site information, collegial interaction, and companionship in the field over the years.

My thanks also go to several anonymous folks from the islands, the family camping at Polihale Beach who shared many interesting facts with us and introduced us to *Waltheria indica* ('uhaloa) and the young woman at the inspection station at the Honolulu International Airport for information on medicinal plants and an interest in this project.

Three friends read parts of the manuscript and provided useful feedback: Rowena Tate, ever sensitive to adverbial abuse; Max Ruedy, representing general interest readers; and Rose Klinkenberg, who provided insights of a field biologist. Many thanks to them all. The entire manuscript was read by Dr. Angela Kay Kepler who made many valuable suggestions for improvements and provided information on sites, situations, and literature of which I was unfamiliar.

A special note of thanks goes to John Alexander of the Dolphin Bay Hotel in Hilo for the safe keeping of the Hawaiian branch of my library, years of excellent service, information on local volcanic events, as well as for providing an endless supply of papayas.

And last, but by no means least, I am indebted to the staff at Mutual Publishing for helping me realize my dream of a book on Hawaiian Island native plants.

Introduction

In 1967, shortly after I had joined the faculty of the Botany Department at the University of British Columbia, I had the pleasure of meeting Prof. Vladimir Krajina who was the senior ecologist at that time. Kraj, as he was known to all, had worked for some time in the Hawaiian Islands making extensive collections of flowering plants and ferns. (One of his early graduate students was Dieter Mueller-Dombois who went on to spend the majority of his very productive career in the Botany Department at the University of Hawai'i.) When Kraj learned that I was planning a trip to the Hawaiian Islands, he took me aside after coffee one morning, and, putting his hand on my shoulder, gave me a "warning." In his most serious voice, he said, "You can not go to Hawai'i only once." Considering that he was one of the most revered members of the senior staff, I was more than a little taken aback by this. But, as it turned out, he was right. As first-time visitors my wife—a botanical artist—and I were overwhelmed by the richness of what we found and knew that there was a lot more we wanted to see. A second trip in 1969 revealed even more wonders. But it was only in the 1970s when I began serious collections in the islands for research purposes that it became obvious that much, perhaps even most, of what we had been seeing wasn't Hawaiian—tropical to be sure—but not Hawaiian. It was those "discoveries" –both the good and the bad—that led in time to the writing of this book. It is hoped that it will help visitors to appreciate the Hawaiian scene: to become acquainted with at least some of the native flora; to recognize some of the aliens and the problems they cause; to respect the work that is being done to reverse some of the harm; and to appreciate how much still remains to be done.

There was another reason behind this book and that is to promote the idea that islands—in addition to being pretty places to visit—are very special, but increasingly threatened, biological laboratories. Biogeographers do not limit themselves to islands in the usual sense—land masses surrounded by water—but extend the definition to include any definable tract of land isolated from its surroundings. Within the context of the Hawaiian Islands, these might include mountain tops, the high volcanoes on the islands of Hawai'i and Maui, for example; soil (edaphic) islands, such as the new lava fields on Hawai'i where growing conditions are different from the immediate surroundings; and patches of forest surrounded by lava (*kipukas*), such as Kipuka Puaulu and Pu'u Huluhulu, both of which occur on the island of Hawai'i (and are mentioned below). This broader definition recognizes the special, and often unique, circumstances associated with the isolation of the plants that grow only on these islands, species that are technically referred to as "insular endemics." Introduction of the concept of non-traditional islands recognizes that it is the isolation itself that provides these unique settings, not the nature of the isolating boundary. A species living in a high elevation environment can be as biologically isolated from its low elevation relatives as any species isolated from its fellows by a thousand kilometers of ocean. Each type of island has its own peculiarities, of course, but each can provide an important, and very valuable "natural laboratory" within which the processes and results of evolution can be appreciated and studied in detail. Organisms, whatever they may be, that have adapted to island life hold important clues on how they deal with the vagaries of Nature. Deciphering these clues requires that we protect the island dwellers from all manner of threats. Perhaps some appreciation for the importance of doing that can be had from the following pages. If we do not protect these precious treasures that occur nowhere else, there may come a time—all too soon, unfortunately—when they will grow no where at all.

Plant geographers deal with three fundamental questions, all of which we will address in this book. As is the case in most scientific work, the basic questions are simple to state; it is the search for answers that can be challenging, or sometimes downright difficult, perhaps even dangerous. The reward in all this lies in the secure knowledge that we will emerge from the exercise having learned just a little bit more about our subject than before we started asking the questions. Another outcome—and this is as close to a guarantee as science allows—is that something totally unexpected is very likely to pop out! That is the nature of science. The questions are: (1) Where is everything? (2) How did it all get where it is? (3) When did it all happen? These topics will appear, in various guises, in the chapters that follow, but let's look at each of these in a bit more detail as they apply to the flora of the Hawaiian Islands.

The answer to the question of where things are, and this isn't limited to islands, has come from generations of people whose goal has been to collect specimens of whatever interesting plant, animal, insect, or mineral they might find on their journeys. Many of these journeys were full-scale expeditions that set out to explore particular areas of interest—the Amazon River basin for example, or the jungles of Borneo, or the foothills of the Himalayas. The other extreme can be as simple as a botanist noticing something unusual about a plant on a local hiking trip. A major scientific effort involving the Hawaiian Islands was the United States

Exploring Expedition of 1838-1842, which brought back numerous specimens from the Hawaiian Islands that were new to science. Although that expedition had natural history collections as a major goal, many others were driven by commercial or strategic interests. Regardless of the primary goal of these voyages, it was a common practice to have a naturalist of some sort on board. In fact, it was not at all uncommon in the days of the great sailing ships for the ship's physician, having studied botany and zoology as part of his medical education, to serve in this capacity. There were commercial incentives for natural history collections as well; collections made in faraway places on behalf of museums could bring a tidy sum to the collector. Some of the early expeditions to the Pacific also included someone skilled in horticulture whose duties involved finding exotic species for display in gardens or for commercial exploitation.

The Scotish botanist David Douglas, the man after whom the Douglas fir was named, and a visitor to Puget Sound and the Fraser River in the Pacific Northwest, was also a major collector in the Hawaiian Islands, having visited on two separate occasions. His first trip was sponsored by the Horticultural Society of London. One of Douglas's biographers described him as "...the most extraordinary and most prolifically successful botanist of all time." Douglas's collecting career came to a disastrous end, however, when he was gored to death by a bull caught in a wild cattle trap on the flanks of Mauna Kea; he was 35 years old. Some mystery surrounds his death, however: an alternative opinion suggests that he was killed by an escaped prisoner and then dumped into the trap. There is a monument to David Douglas on the northeastern flank of Mauna Kea.

A generation earlier saw three voyages by Captain James Cook, who visited the Antarctic Ocean, the coast of New Zealand, the New Hebrides, discovered New Caledonia, discovered the Hawaiian Islands (which he named the Sandwich Islands after his patron the Earl of Sandwich) in 1778, and tried, unsuccessfully, to find a passage from the Northeastern Pacific Ocean to the Atlantic. Joseph Banks (later Sir Joseph), an English naturalist, accompanied Cook on his first journeys and made significant collections of plants not known to science at that time. The genus *Banksia*, one of the spectacular proteas first observed on those trips, is named in his honor. After his first failed attempts to locate the passage to the Atlantic, Cook's intentions were to spend the winter in the warmer climes of his newly discovered Sandwich Islands and continue the quest for the passage the following year. In early February, 1779, while attempting to repair one of *H.M.S. Resolution's* masts, Cook and his crew met with some hostility in interactions with natives at Kealakekua Bay on the leeward coast of the island of Hawai'i. Several incidents involving theft of tools from his ships resulted in a tense, if not outright dangerous, situation. Indeed, serious resistance was met when Cook and his men attempted to recover a stolen cutter. A fight ensued on the beach, and he was killed on the morning of February 14, 1779. Present at this fateful altercation was King Kamehameha I, who was the first to unite the Hawaiian Islands under a single government. We will meet him again.

Other visits to the Hawaiian Islands of historical interest include Archibald Menzies, surgeon and naturalist on the H.M.S. *Discovery* under Captain George Vancouver, who visited the islands in 1792, 1793, and 1794. James Macrae, a Scottish botanist on the H.M.S. *Blonde*, under Captain George Anson (Lord Byron, cousin of the poet), visited the islands in 1825. Macrae may have been one of the first Europeans to see the silversword plants (*Argyroxiphium sandwicense*) in Nature. William Rich was the botanist, Charles Pickering and Titian R. Peale the naturalists, and William D. Brackenridge the horticulturist on the United States Exploring Expedition of 1838-1842 under Commodore Charles Wilkes, who, incidentally, was not a very nice person at all, but more on that later. William Hillebrand, a medical doctor, collected widely during the period 1851-1871 and published the *Flora of the Hawaiian Islands* in 1888. Several botanists, working either at the Bishop Museum or at the University of Hawai'i, both in Honolulu, have made impressive contributions to our knowledge of the islands' flora. Charles Noyes Forbes, also associated with the Bishop Museum, collected over 9,000 specimens during the period 1908-1920. Joseph Francis Charles Rock, over a long career teaching in the islands, collected over 30,000 specimens! Equally impressive were the accomplishments of Otto Degener, who during the period 1922-1988, also amassed a collection of over 30,000 specimens. Over a span of many years, Degener published a collection of descriptions of plant groups under the title *Flora Hawaiiensis*. Carl Johann Frederik Skottsberg, director of the botanical garden and professor of botany at the University of Göteborg in Sweden, collected over 2,000 specimens and wrote about plant colonization of lava. One last contributor, and a name that is synonymous with Hawaiian botany, was Harold St. John whose activities in the islands, from 1929-2000, resulted in over 9,000 specimens being deposited in the collection at the Bishop Museum. To be sure, many contemporary botanists continue to make important contributions to our understanding of the nature and origin of the islands' flora, but attempting to list them would undoubtedly result in the omission of key players. Suffice it to say that important contributions from many of them appear in the pages that follow.

Upon returning to their home ports, collectors would give, or—if under contract—sell, their specimens to some university or museum where authorities could study, name, and categorize them. Specimens were then deposited in permanent collections for future reference. Over time, some of these collections grew quite large, numbering well into the millions. For example, the collections at Harvard

University, housed in the Gray Herbarium, where many plants collected on the United States Exploring Expedition reside, has 4.6 million "sheets," as individual specimens are known. Other large collections—all counting in the millions—include those at the New York Botanical Garden, the Royal Botanic Garden at Kew, the British Museum, Paris, and Leningrad. In the days of Darwin, Douglas, and others, most of the active botanical study was done in Europe where extraordinarily important collections were maintained, hence the last four listed institutions. Not surprisingly, the Bishop Museum in Honolulu also houses a significant collection—over 550,000 specimens, including, of course, many collected in Hawaiian Islands.

Field collections continue today, of course, although much of the work lacks the glamour of adventures on the high seas that seemed to characterize the "good old days." There are numerous botanical journals active today devoted to the description of new species based on specimens collected in all parts of the world. Specimens continue to be sought, especially in areas of particular botanical interest that were not accessible during earlier expeditions owing to extremes in terrain or hostile local inhabitants. Technology has also come to the collector's rescue: for those who can afford it, there is the helicopter! Helicopter access to remote mountainous areas in Hawai'i has been of immense help to botanists in the islands. Other botanical adventures in the Hawaiian Islands have required the use of mountain climbing techniques to explore the severe cliff faces of several of the islands.

Another problem confronting field biologists is seasonality; collectors may not be in a particular area at the right time. [In preparing a guide book to Kakadu National Park in Northern Territory, Australia, a colleague of mine, Matthias Breiter, photographed at least one local denizen long considered extinct. He was simply there at the right time and found the beast present in some number.] There are also species of desert ephemerals native to the arid regions of southern California and Arizona that only appear if the growing conditions—the correct amount of rainfall at the proper time—are met.

Although our focus in this book will be on the question of how native species on islands become established, it is useful to look briefly at the other extreme of plant distribution. A few plant species occur so widely that they are described as cosmopolitan in distribution. Perhaps the most recognizable species in this category is the bracken fern (*Pteridium aquilinum*). Bracken is known on all continents, except Antarctica, and often occurs in dense stands sometimes spreading over entire hillsides; see Plate (117) for an example in the Hawaiian Islands. Some workers prefer to recognize bracken as consisting of several species. Under that treatment, the original species would lose its cosmopolitan status, but the genus, *Pteridium*, could then be described as a cosmopolitan genus. Dandelion (*Taraxacum officinale*), likely among

Plate 1. The endemic *Tetramolopium humile* and an alien species of *Hypochaeris* sharing a niche on Haleakalā.

the best known plants in the world, is considered to be a cosmopolitan weed. The last word, "weed," is key in that description: weedy species are generally excellent long distance travelers and can usually take advantage of almost any opportunity to become established. An example of a weedy species in the Hawaiian Islands is a lawn weed common in North America, a species of *Hypochaeris* (Plate 1), a close relative of the dandelion. In this picture the weed is sharing a rocky crevice with the Hawaiian native species *Tetramolopium humile*. Owing to the vulnerability of island ecosystems, agriculture and commercial development provide many excellent opportunities for invasive weedy species. With their warm climate, lack of frost except at higher elevations, ample rainfall, and nutritionally rich soils, the Hawaiian Islands offer particularly attractive opportunities for the establishment of weedy species. We will look at the very serious problem of invasive species in Chapter Three.

This brings us to question number two, how do closely related species find themselves separated by distances of hundreds or thousands of kilometers? There are two ways that distributions involving such distances can occur. The first of these, called dispersal, involves movement of some plant part; seeds are obvious candidates but any vegetative part capable of setting roots (think of rooting geraniums) will work, or spores in the case of ferns, fern allies, and mosses. Intimately associated with this manner of populating an island is the nature of the propagule itself; in other words, what special characteristics must it have to make a long journey and survive in its new habitat? For oceanic islands the main dispersal mechanisms have been via air currents, flotation, birds, and possibly by rafting on logs.

The other means by which plants travel over significant distances, or at least appear to have traveled over long distances, is termed vicariance and involves geological events that result in entire land masses splitting apart and their separate parts rafting in opposite directions. This process does not single out individual species of plant for transport; rather, entire tracts of land are involved that carry with them everything that grows there. This is a process that happens on a grand scale, such as the ancient rafting apart of Africa and South America, the separation of Europe from North America, the movement of Madagascar (and India) away from the African continent, or the present day disruption of Iceland caused by seafloor spreading along the mid-Atlantic Ridge. Although our concerns with getting plants to islands will deal exclusively with dispersal, the tectonic movement of plates across great distances on the face of the planet has profound influence on how oceanic islands arise in the first place. We will look at these processes in Chapter One.

The third question, which has to do with when a particular plant distribution might have become established, requires an appreciation of geological time. When we deal with such topics as extended periods of volcanism or the birth and death of islands, we are not dealing with time in terms of human life-times, or even the lifetimes of civilizations; instead, we must try to push our imagination literally millions of years into the past. In truth, the times can be mind boggling. For example, the first primitive land plants are thought to have lived about 430 million years ago. Extensive growths of ferns, giant horsetails, and cycads, plants that were to become massive coal deposits, lived in swampy lowlands from about 360 million to about 285 million years ago in a time called the Carboniferous Period. Thanks to Hollywood, probably the most familiar term dealing with biological history is the Jurassic Period, which spanned the period between about 200 million and about 145 million years ago. We have come to know this period as the age of the dinosaurs. The first fossil record of flowering plants dates to about 135 million years ago, and many modern families of flowering plants can be recognized in Northern Hemisphere fossil beds that are about 95 million years old. Pollen of the sunflower family (Asteraceae) showed up in the fossil record about 30 million years ago and was found in abundance in deposits worldwide soon thereafter indicating its exceptional capacity for long distance travel. The genus *Homo*, of which we are a member, first appeared a few million years ago, *Homo sapiens*, our species, perhaps a scant two hundred thousand years ago. At that time, the current Hawaiian Islands would have been easily recognizable, with the youngest of them, Hawai'i, in an early stage of fiery growth.

The present high islands in the Hawaiian archipelago range in age from about five million years for Kaua'i, to about a half million years for Hawai'i. I should emphasize that these ages represent the oldest known rocks on those islands. For

Hawai'i, the oldest rocks occur on Mt. Kohala at the northwestern corner of the island, while the youngest are being formed through volcanic activity at the present time on or near the eastern coast. I like to think of the age of Hawai'i as starting from zero, the moment that fresh lava cools enough to support my weight. In Chapter One we will see the progression of ages of the islands that make up the main group of high islands as well as the northwestern extension of the chain that extends to Midway Island and Kure Atoll. We will also consider very briefly the Emperor Chain, seamounts that extend even farther into the northern Pacific Ocean.

Returning to the sequence of historical events that bring us more or less up to date, we now know that the Hawaiian Islands were originally populated by two waves of colonists, the first from about 100 B.C. to 200 A.D. from the Marquesas, and the second in the twelfth to fourteenth centuries from the Society Islands (Ziegler, 2002). The first visit to the Hawaiian Islands by Europeans is credited to Captain James Cook and his crew who first saw Kaua'i on January 18, 1778. He was killed about a year later. This first meeting is referred to as the "Contact." Dedicated settlement of the islands by non-native peoples began in earnest in the early 1800s.

The Hawaiian Islands existed as a monarchy until 1893 when Queen Liliuokalani was overthrown. Efforts to re-establish the monarchy failed, and in 1894 the islands became the Republic of Hawai'i. Annexation to the United States was proclaimed on August 12, 1898, with full territorial status granted in 1900. Statehood was proposed in 1937 but the request was refused by the United States Congress. Two reasons were given for the decision, the obvious one being the distances involved, the other a reflection of the xenophobia that characterized the United States at that time—there were too many "non-Americans" on the islands. The distances shrank as air travel became more accessible, and the attitude toward the many nationalities present on the islands changed to one of recognition of the cultural richness brought by the multitude of backgrounds they represented. Hawai'i became the 50th member of the United States in 1959.

Before embarking on our tour of the islands and their unique plants, it is necessary to define several other terms and concepts that will be encountered in this book. The first one is the word "island" itself. When speaking about islands in the traditional sense—a body of land surrounded by water—it is necessary to understand that there are two fundamental types of islands, continental and oceanic. A continental island is simply an island that lies close to a continent, is separated from it by a comparatively shallow channel, and is part of the continental plate itself. During periods of lowered sea level the island was connected to the larger land mass. Familiar examples include the British Isles, which were part of the European continent, and the islands in Puget Sound, those in the Strait of Georgia, and Vancouver

Island itself, which were at one time part of the North American continent. During the period when the larger and smaller land masses were connected a free flow of plants and animals could occur. Because there were essentially no barriers to movement of either plant propagules or animals, the smaller and larger areas would, therefore, have been home to much the same biota. This situation would have been maintained after the island had become separated from the mainland by rising sea level. Isolation of the island for a lengthy period of time could, of course, result in differentiation of some elements of the biota such that new varieties or subspecies might be recognized.

The second kind of islands are volcanic in origin. They occur for the most part in deep oceans and have never been attached to any continental land mass. They start out life as sterile mountains of lava, acquire a variety of habitats as they mature, suffer the erosive effects of wind and water, waste away to mere remnants, and eventually disappear beneath the sea. During their younger years they acquire their residents piece by piece. In the case of the Hawaiian Islands, it has been estimated that their flora is the result of about 280 colonizations. Averaging this number of events over the age of the islands, it is possible to argue that one colonization every 20 thousand years or so could account for the original flora. Although I have seen larger estimates, the fundamental point here is that colonization of the islands was a slow process. Considering the luxuriant vegetation on the islands now, it is difficult to imagine a new island with only a few scattered plants, but that certainly must have been the case. As a new lava island continues to grow, as noted above, a variety of habitats are formed, some at low elevation, some higher up the mountain, some on the windward and others on the lee side of the island. The eventual colonization of the island by plants, which leads to accumulation of organic debris, as well as continual weathering of the original rocks contribute additional new habitats.

The availability of a variety of habitats, however, is just one part of the picture. In order for successful colonization of a flowering plant to occur, certain requirements of its seeds must be met. Various adaptations that provide seeds with the capacity for long distance dispersal include: size; the presence of some structure that gives it air worthiness, such as the feathery device used by dandelions; the presence of hooks, barbs, or a sticky substance that would allow it to adhere to a bird; an attractive and tasty fruit that could ensure it being carried inside a bird; or some kind of means of flotation, such as a thick husk seen in the coconut. There are risks associated with long distance dispersal by air, however, desiccation and the sensitivity of the seed to cold temperatures and UV radiation in the atmosphere, or if by flotation, the need to survive long immersion in seawater and remain viable.

Another factor that comes into play in island colonization is the breeding biology of the plant. Outcrossing, the exchange of genetic material between different plants, is an adaptive strategy that helps to maintain a high level of genetic variation within a species. Outcrossing minimizes the accumulation of potentially deleterious genes, a process that tends to occur in self fertilization. Ways in which outcrossing is assured include incompatibility reactions in which a plant rejects its own pollen, or having male and female flowers on different plants, or requiring a specific insect pollinator, or, if male and female flowers exist on a single plant, one of the sexes matures before the other. It is easy to see that in the first two cases at least that a single individual may be in trouble. A seed of a self-incompatible plant may get to an island, land in a favorable habitat, germinate, and grow, but if it cannot accept its own pollen it is doomed. There are ways around the problem, however. If the plant happens to be a perennial, or can become perennial, or if it is an annual with the capacity for vegetative growth, it may survive in its new home. Successful sexual reproduction, however, would require the arrival of a compatible mate, an event that, while possible, has a low probability of happening. Given sufficient time, however, a self-incompatible colonist may undergo a mutation that eliminates the incompatibility reaction. This enables the species to expand its range in the new habitat and in time may experience a return to incompatibility. Such flip-flops are a well-known property of some plants. Since many island species have been shown to be outcrossers, or to have sexes on different plants, it is clear that these systems are fairly flexible.

To this point we have been talking exclusively about flowering plants, but a major contribution to floras comes from ferns and other spore-bearing plants, e.g., mosses and horsetails. It is common for these plants to make up a larger proportion of an island's flora than would be expected based upon their contribution to their home continent's flora. Warren Herb Wagner (1995) pointed out that the ratio of native ferns to native flowering plants in the Hawaiian Islands is roughly 1:6, whereas the ratio is closer to 1:14 for continental floras. Ferns are often among the first invaders of new lava on the Hawaiian Islands and were among the first plants to colonize Rakata, the new volcanic island formed after the eruption of Krakatau. In August of 1883 Krakatau Island, which lay in the Sunda Strait between Java and Sumatra, exploded in what has been judged to be the largest volcanic eruption in historic times (Thornton, 1996; Winchester, 2003). Reborn after the massive destruction of the original islands was Rakata. The first botanist to set foot on Rakata was M. Treub who was part of a party who visited the island for four days in June of 1886. Away from the actual coast Treub found 11 species of ferns, 15 species of flowering plants, and two species of mosses. The ratio of fern species to flowering plant species is, roughly, 1:1.5, clearly indicating the superior colonization capacities of spore-bearing plants. Subsequent visits to Rakata resulted in collections with increasing ratios of flowering plants but the numbers never reached continental values. I should note that while some ferns are good colonizers, they are not good competitors once flowering plants begin to claim space. It should not be surprising that most of the flowering plants found on Rakata were those normally found on beaches, e.g., beach morning glory (*Ipomoea pes-caprae*) and a species of *Scaevola* (called *naupaka* in the Hawaiian Islands)—we'll meet these later on in this book—along with others. All of this suggests that there may be something about ferns that make them particularly good colonizers. Is this true? The answer is yes, and it has to do with the way they reproduce.

Spore-bearing plants have a major advantage over seed plants in that their propagules, the spores, are extremely well adapted for long distance travel. They are much more resistant to desiccation than most seeds and can lie dormant in the soil for long periods of time only germinating when conditions are right. Seeds, which consist of an embryo and often some food reserves packaged in a comparatively thin coat, can be quite large by comparison. In contrast, spores are very small with their contents enclosed by an exceedingly tough covering. For simplicity's sake we need only think of spores as containing the minimal amount of information needed to

Plate 2. *Polypodium pellucidum* var. *vulcanicum.*

produce a new plant called a gametophyte (literally, sex cell producing plant). It is the function of gametophytes, which tend to be small and inconspicuous, to produce male and female reproductive cells (the gametes). Union of male and female gametes results in the formation of the familiar fern plant in its sporophyte stage (literally, spore-bearing plant). Spores are produced in structures on the undersides of fern leaves. These brownish structures, called sori (singular, sorus), can be seen in the photograph of *Polypodium pellucidum* var. *vulcanicum* (Plate 2).

Many ferns produce only one kind of spore (called homosporous), but there are others that produce two kinds, large ones and small ones (called heterosporous). The large and small spores develop into separate gametophytes, one of which produces male gametes while the other produces female gametes. As before, fusion of these gametes results in a new sporophyte. This mode of reproduction, which requires two plants, is similar to the situation of an outcrossing seed plant where two individuals are required. Needless to say, ferns with this mode of reproduction are not as abundant on islands as those that utilize the single spore strategy. The larger majority of Hawaiian ferns are homosporous; only the water fern *Azolla* among the Hawaiian ferns is heterosporous.

Another feature of plant life is that many higher plants live in close symbiotic relationships with soil fungi. These relationships are of particular importance to plants that live in nutrient poor soils, such as on beaches, where the fungi assist in procurement of nutrients for their host plants. An examination by Richard Koske of the University of Rhode Island and J. N. Gemma of the National Tropical Botanical Garden (1990) revealed that many plants on the Hawaiian Islands also have fungal associations. This naturally led to the question of how two members of such an association could have got there, separately or together? In fact, it appears that 'both ways' would be the correct answer. Fungal spores can travel as baggage on plant material washed ashore—they retain some viability after 12 days in dilute salt solution—and could conceivably adhere to a seed being carried externally by a bird. And, since fungal spores are very tiny, they can easily be transported to essentially anywhere on the Earth via air currents.

A uniform code of plant nomenclature has been agreed upon that allows for efficient communication between botanists. We will use parts of the system here. The system is hierarchical with categories that range from kingdom, e.g., plants and animals, at the top end, to form (formae), which exhibit relatively minor differences between individuals at the other end. The units that will be used in this book are mainly species, genus, and family. A species consists of a number of individuals that share a particular suite of characters, a genus consists of one or more species, and a family of one or more genera. Patterns of variation within each of these ranks can also be recognized formally by use of subgenus, or subfamily, a few of which will appear in the text. It is common to recognize lesser differences within species by defining either varieties or subspecies. Many botanists prefer to limit the subspecies category for two sets of populations within a species that are separated geographically. An example of this usage will be described in Chapter Two using the silverswords as an example. Placing a given set of organisms into these categories is very much a matter of opinion and not everyone agrees with these assignments. The advantage of working with a uniform set of standards is that you always know what someone is talking about, even if you do not agree with them. The term "taxon" (plural, taxa) will be encountered from time to time in this book as well. Taxon refers to the members of some formal rank, species or genus for example, and is used as a general term once the object group has been identified.

The term "population" will be used from time to time in this book. A population is <u>not</u> a category in the formal hierarchy, but the concept is nonetheless of very real significance since it generally refers to a collection of individuals within a species that comprise a breeding unit. Thus, populations can be thought of as assemblages of functioning, interbreeding organisms, whereas the other terms refer strictly to a formal system of naming and classification.

A variety of terms also exists that deal with environmental variation, the most important distinction of which, for out purposes, has to do with available moisture. The term "xeric" (literally, dry) refers to desert-like sites. At the other extreme is the term "hydric" referring to aquatic environments, which we will not deal with in this book. Between these extremes we use the term "mesic," which means, logically enough, middle. Remember that this is a continuous scale and that, for our purposes, an overall sense of the environment is all that is , e.g., "subalpine xeric shrubland," or "mesic to wet forest."

A Short Library of Hawaiiana

I have been fortunate in having had the opportunity to visit and work in the Hawaiian Islands off and on for nearly 35 years, and I hope to share with the reader some of the things I have seen over that time. However, in the preparation of a book such as this one, an author turns to many others for additional information, background, and insights. Some of these major works are outlined briefly below. Full references can be found in the Bibliography where citations of studies from the research literature also occur. The principal source of botanical information for the Hawaiian Islands is the two volume *Manual of the Flowering Plants of Hawai'i* edited by W. L. Wagner, D. R. Herbst, and S. H. Sohmer. It first appeared in 1990, with an updated version published in 1999 (University of Hawai'i Press). This is a detailed, technical flora that is well illustrated with line drawings. The first 120 pages or so contain descriptions of geography, climate, and vegetation, along with a history of botanical collection in the islands. I will refer to it throughout this book as the *Manual*. The ferns of Hawai'i have been described in a technical flora by D. D. Palmer (2003) and in a more general manner by K. Valier (1995).

Less technical books, uniformly well illustrated with color, include *Plants and Flowers of Hawai'i* by S. H. Sohmer and R. Gustafson (1987), *Remains of a Rainbow: Rare Plants and Animals of Hawai'i* by D. Liittschwager and S. Middleton

(2001), *Trailside Plants of Hawai'i's National Parks* by C. Lamoureux (1976), S. H. Lamb's (1981) *Native Trees and Shrubs of the Hawaiian Islands*, A. K. Kepler's (1990) *Trees of Hawai'i*, and W. A. Whistler's (1980) *Coastal Flowers of the Tropical Pacific*. A *Pocket Guide to Hawai'i's Trees and Shrubs* by H. D. Pratt (1998) is highly recommended. Also available is a series of books devoted to individual islands by A. K. Kepler: *Haleakalā: A Guide to the Mountain* (1988), *Majestic Moloka'i* (1991), *Maui's Floral Splendor* (1995), and *Hawai'i's Floral Splendor* (1997),

For an overview of the natural history of the Hawaiian Islands, there is no better source than *Hawai'i. A Natural History* by Sherwin Carlquist (1980) and its companion *Island Biology* (1972). Two recent publications of note are *Isles of Refuge* by M. J. Rauzon (2002), which deals with the northwestern islands, and *Hawaiian Natural History, Ecology, and Evolution* by A. C. Ziegler (2002). The 3rd edition of *Atlas of Hawai'i*, edited by S. P. and J. O. Juvik (1998), is a beautifully illustrated and valuable source of information on many aspects of Hawaiian life.

Excellent sources of ethnobotanical, including medicinal, uses of Hawaiian plants, along with frequent references to folklore underlying the uses, can be found in *Native Planters in Old Hawaii* by E. S. C. Handy et al. (1992), *Lā'au Hawai'i: Traditional Uses of Plants* by I. A. Abbott (1992), *Plants in Hawaiian Culture* by B. H. Krauss (1993), *Hawaiian Heritage Plants* by A. K. Kepler (1998), and *Hawaiian Herbal Medicine. Kāhuna Lā'au Lapa'au* by J. Gutmanis (2001). Growing Hawaiian plants is described in detail by J. L. Culliney and B. P. Koebele (1999). A classic work describing native and cultivated plants is M. C. Neal's *In Gardens of Hawaii* (1965).

Hawaiian names will be given in many places throughout this book. These follow usage in the *Manual* but I have also consulted the Hawaiian Dictionary by M. K. Pukui and S. H. Elbert (1986) which source is referred to as *P. & E.*

The history of the Hawaiian Islands from the time of the Cook expeditions to modern times, including often Byzantine political intrigue, is thoroughly covered by Gavin Daws in *Shoal of Time. A History of the Hawaiian Islands* (1968).

Highly recommended for a colorful survey of the islands' flora is the beautiful display of Hawaiian plants available through an award winning web-site prepared by Prof. Gerry Carr of the University of Hawai'i. The address is: www.botany.Hawaii.edu/faculty/carr/native.htm. Thumbnail pictures lead to full screen displays. There are also links to related topics, such as a detailed look at the silversword alliance, and a list of alien plants on the islands. An important source of additional information on rare and endangered species of Hawaiian vascular plants (flowering plants and ferns) can be found in an article by F. R. Fosberg and D. Herbst published in 1975.

CHAPTER ONE

The Islands

As commonly described, the Hawaiian Islands consist of eight "high" islands, islands that have their highest elevation greater than a few hundred meters. From east to west the islands are Hawai'i, Maui, Kaho'olawe, Moloka'i, Lāna'i, O'ahu, Kaua'i, and Ni'ihau (Figure 1). With the exception of Kaho'olawe, which is uninhabited, and Ni'ihau, which is off-limits to non-Hawaiians, these are the islands visited by tourists. To recognize the full extent of the Hawaiian chain, however, we must take into consideration the many islands that lie to the northwest of Kaua'i, stretching about 1,900 km to Midway Island and to Kure Atoll, another 100 km or so. The chain continues beyond Kure to the Daikakuji Seamount (ca. 3,500 km northwest of

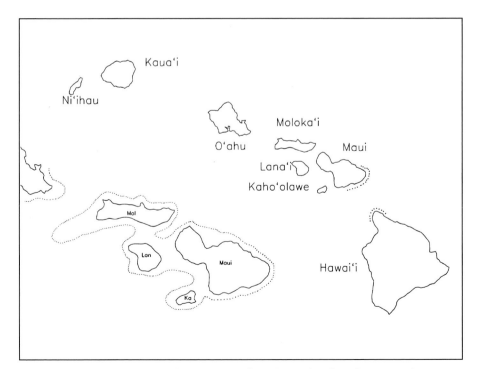

Fig. 1. Map of the main Hawaiian Islands. The inset shows the rough outline of Maui Nui. The Penguin Banks extend to the west-southwest from Moloka'i.

Kaua'i) where it makes an abrupt turn to the north (the Pacific Plate changed direction). From that point the chain continues to Meiji Seamount, which lies between the end of the Aleutian Chain and the Kamchatka Peninsula. We will look at each of the major Hawaiian Islands in turn, pointing out how they came into existence, and some of the important features of each that figure in the discussions of plant groups later on. Next will be a brief visit to a few of the islands that comprise the northwestern extension of the Hawaiian Islands, and finally, a few comments will be made on the existence of even older islands, those belonging to the Emperor Seamounts.

Much of the geological information on which this chapter is based comes from the lavishly illustrated *Volcanoes in the Sea. The Geology of Hawaii* by G. A. Macdonald, A. T. Abbott, and F. L. Peterson (1983), a chapter by H. L. Carson and D. A. Clague on geology and biogeography of the islands in *Hawaiian Biogeography, Evolution on a Hot Spot Archipelago* edited by W. L. Wagner and V. A. Funk (1995), and the recently published *Hawaiian Natural History, Ecology, and Evolution* by Alan C. Ziegler (2002). Elevations in meters are taken from the chapter on geology and biogeography of the islands by Carson and Clague. Elevations in feet are taken from the maps of the Hawaiian Islands prepared by J. A. Bier (University of Hawai'i Press). Ages of the northwestern islands are those used by Price and Clague (2002).

In a nutshell, the Hawaiian Islands have come into existence through the existence of a "hot spot" on the floor of the ocean over which the massive Pacific Plate has moved—in a northwesterly direction—leaving a string of mountains in its path (Wilson, 1973). Massive amounts of molten rock—called magma, until it bursts forth on the surface where it becomes lava—accumulate and become sea mounts. Those that break the surface in this region of the Pacific are called the Hawaiian Islands. Because the Pacific Plate continues to move, each newly formed island will in time be left behind as the next is beginning to be formed. The fate of each island is assured from the day of its emergence: growth and maturation, loss of material through erosion by wind and water, possibly a brief rejuvenative eruption or two, and eventual disappearance. The overall process takes several millions of years, depending upon the size of the original accumulation. (As we will see below, Kaua'i is about 5.1 million years old and as such is the oldest of the current high islands.) It is during this period of time that the individual islands undergo development of a complex array of potential niches—growing sites for colonizers, undergo colonization following the chance arrival of seeds or other biologically viable materials, mature into complex biological communities, and undergo eventual decline as erosive forces remove more and more of the islands. A typical visit to the Hawaiian Islands could include a visit to the Island of Hawai'i to see an active volcano and a visit to the northwestern corner of Kaua'i to stand on the oldest rocks currently available in the islands.

Before visiting the individual islands, it is useful to contemplate for a moment or two just how massive some of these oceanic mountains are or might have been. We can start by looking at Mauna

Plate 3. View of Mauna Loa from the Jaggar Museum parking lot. The long, gentle slope is typical of a shield volcano.

Loa, the southernmost active volcano on the Island of Hawai'i . This giant rises to 4,169 m (13,796 ft.) above sea level making it significantly taller than Mt. Everest, when measured from the ocean bottom. The shear mass of this colossal accumulation of rock is not immediately apparent when one looks at it from the Devastation Trail, for example, or from the parking lot of the Volcano Observatory (Jaggar Museum) (Plate 3). The reason for this deception is the very gentle angle of its slope, which is characteristic of shield volcanoes. An even larger member of this group of giants was thought to have existed where we now find Haleakalā National Park. It has been estimated that when this mountain was at its maximum height, it would have risen to about 5,000 m (ca. 16,400 ft)!

Plate 4. A view into Kīlauea Crater showing Halema'uma'u fire pit.

Plate 5. Lava flowing into the ocean near the vicinity of the former town of Kalapana, Hawai'i.

One of the most impressive features of the Big Island—and most often visited—is the ease with which it is possible to examine the activities of an active volcano. The caldera at Kīlauea Volcano is particularly easy to see since it is one of the major attractions on the Ring Road in Hawai'i Volcanoes National Park. A walkway leads to the rim of the caldera within which one can see the Halema'uma'u fire-pit (Plate 4). Refilling of this pit by fresh lava followed by repeated collapses were common events during the history of the volcano and have been tracked in detail by scientists working at the Volcano Research Center. Filling and collapse cycles are not likely to be as common as they were in the past, however, since much of the pressure beneath the volcano is being released through continued eruptive events farther to the east. The results of these eruptions (details below) can be seen by following the Chain of Craters Road eastward to the coast. One clearly sees evidence of recent volcanic activity along the road with several of the older craters clearly named with their dates of birth indicated. The road, which used to continue north along the coast, has been closed for several years owing to the continuing lava flows along the southeast rift zone. A popular activity is to visit the lava flows at night, either from the route just described, or from the north following Rt. 130. Lava flows at night, whether flowing over older lava or into the ocean, can be quite striking (and warm!) as can be seen in Plate (5).

Despite their size and impressive activities, volcanoes do not last forever. Staying with Haleakalā as our example, it is safe to say that it isn't the volcano it used to be. At the present time its highest point is 3,055 m above sea level, almost 2,000 m less than it is thought to have been at its greatest height. Since Haleakalā has been shown to be only about 750,000 years old, it is clear that its losses over this period of time have been substantial. What can bring about degradation of this magnitude? Collapse of the original top of a volcano to form its caldera is of course one of the things that can happen, but subsidence, continual action of crashing surf, mass loss through landslides, and

Plate 6. An example of erosion caused by water near Rt. 550 on Kauaʻi.

extensive erosion from millennia of heavy rainfall, also take their toll. Plate (6) shows the effects of water erosion on a hillside near the junction of Routes 550 and 552 on Kauaʻi.

But volcanoes at this stage of their lives still have something to say! Despite the absence of eruptive activity for long periods of time, one is cautioned to consider these giants as dormant, rather than extinct. Post erosional eruptions, the rejuvenation stage of a volcano's life, have been common events on most of the Hawaiian archipelago. Haleakalā, for example, was last heard from in about 1790—the lava flow from this eruption formed Cape Kīnaʻu on the southwestern tip of East Maui, not far from the Mākena Golf Course. The story of how the date of this eruption was established has been told by Alan Ziegler (2002) in his recent book on natural history, ecology, and evolution in the islands. The story is worth a brief retelling here. The southwestern coast of Maui was mapped in 1786 by Captain Jean-François G. de La Pérouse, a French explorer. The area was again visited and mapped in 1793, this time by Captain George Vancouver. Comparison of the two maps revealed that Cape Kīnaʻu exists on the more recent map but not on the earlier one, hence the 1790 date. Captain de La Pérouse's explorations have been recognized by the naming of several sites after him, among which are La Pérouse Bay, which lies just to the south of Cape Kīnaʻu, the Mākena-La Pérouse Bay State Park just north of the cape, and one of the northwestern islands, La Pérouse Pinnacle.

Plate 7. Coral reef at Kē'ē Beach on the northern coast of Kaua'i.

Plate 8. A view of the Nā Pali Coast, Kaua'i, looking west from near the trailhead of the Kalalau Foot Trail.

One of the features of aging volcanic islands is the development of coral reefs. One of the easiest coral reefs to see in the Hawaiian Islands can be found at Kē'ē Beach at the northwestern end of Rt. 560, which is an extension of Kaua'i Rt. 56. This is a favorite spot for snorkeling as well as for enjoying spectacular sunsets. An excellent view of the reef can be had by climbing a short distance up the trail to Hanakāpī'ai Beach. Incidentally, this is also one of the spots most visited by photographers who wish a view of the shear sea cliffs that define Kaua'i's northwest coast. See Plates (7) and (8), respectively, for views of the coral reef and the cliffs.

Island of Hawai'i—The Orchid Isle

We will start our island hopping on the Island of Hawai'i, nicknamed the Orchid Isle, but most commonly referred to as the Big Island. The island is built upon six volcanoes, two of which are still active. (Whether Lo'ihi eventually joins with Hawai'i remains to be seen.) The oldest of the six volcanoes was Māhukona, which never achieved an elevation of any great significance (ca. 235 m) and is today over a thousand meters beneath sea level. The site of this volcano is marked by the present town of Māhukona and by a beach park of the same name. Eruptions from the next volcano totally over-topped Māhukona.

The oldest major volcano on Hawai'i was Kohala, the remains of which make up the northern tip of the island. Mt. Kohala currently stands 1,670 m above sea level, a loss of 1,000 m from its estimated highest elevation. This dormant volcano has been dated at 430,000 years. The western flank of Mt. Kohala falls away gently to the coast, whereas the northeastern flank shows the steep sea cliffs typical of the windward coasts of islands that have been at the mercy of both crashing surf and ground water erosion. The effects of these actions can be appreciated by visiting Waipi'o Valley or by continuing to the northern tip of the island and following Rt. 270 to its termination at the Pololū Valley Lookout. A trail leads to the beach with views of both the valley and of rugged sea cliffs (Plate 9).

Plate 9. Sea cliffs of northeastern Hawai'i viewed from the Pololū Valley lookout.

One of the great treats awaiting visitors to Hawai'i is the opportunity to visit the top of one of the largest dormant volcanoes on the Earth with comparative ease (and safety). There are restrictions on the type of vehicle allowed on the summit road (four-wheel drive required), but this does not deter the many thousands who visit Mauna Kea each year. Rising to a breathtaking (literally) height of 4,205 m (13,796 ft), the trip to the summit of this 380,000 year-old giant provides magnificent views and the chance to visit one of the most advanced astronomical observation facilities on Earth, where an array of telescopes is maintained by astronomers from Canada, France, Great Britain, Japan, the Netherlands, and the United States (Cruickshank, 1986). Plates (10a and 10b) show several of the facilities.

Plate 10a. Astronomical observatories on the summit of Mauna Kea.

Plate 10b. Astronomical observatory on one of the lower summit sites of Mauna Kea. Note Mauna Loa in the distance.

This is a convenient point to consider some extremes. In the early morning, when the weather is clear in Hilo, it is possible to see sunlight reflected from the observatories on top of Mauna Kea, or reflected from the snow that accumulates at that elevation in the cooler months. The distance is deceptive; the domes appear to be just up the hill a bit. By line of sight they are about 30 miles away, without taking into consideration the trigonometric correction for the summit's nearly 14,000 ft elevation. A view of Mauna Kea including the coastal town of Hilo can be seen in Plate (11). It is difficult to imagine that the difference in elevation represented in this photograph—sea level for the photographer, almost 14,000 feet for the summit—is over two and a half miles. As we saw above with Mauna Loa, the slope of a shield volcano is so gentle as to conceal its impressive vertical dimensions.

Although it may have been clear in the early morning, chances are very good that it will not stay that way; located on the eastern coast of the island, Hilo can receive almost daily rainfall when the northeast trade winds blow. The trip to

Plate 11. Mauna Kea in the distance with the town of Hilo across the bay.

the summit involves driving on Route 200 to the saddle—it's called the Saddle Road—that lies between Mauna Kea to the north and Mauna Loa to the south. At about the 7,000 ft level the summit access road takes off to the north, and continues up the mountain (remember, ordinary cars are not allowed past the Visitor's Center). A visit to the Ranger Station is recommended where valuable advice is available. As one gains elevation, the clouds thin, and by the time one reaches the summit, they have usually been left well below. Among other things, the trip to the summit allows the visitor to appreciate the immense size of a shield volcano and the significant effect that a land form of this dimension has on local weather. Hilo lies in the tropical rainfall zone and has the lush vegetation to show for it, whereas the top of Mauna Kea is an alpine desert with little capable of growing under such extreme conditions. This gradual but profound change in conditions from sea level to summit is one of the factors that provides the many microhabitats that are home to many of the unique plants, animals, and insects of the islands.

The changes in vegetation patterns that occur as one travels from the sea shore to the mountain tops have been recognized by identifying five principal plant communities based on elevation: (1) coastal, 0-300 m; (2) lowland, 15-2,000 m; (3) montane, 500-2,700 m; subalpine, 1,700-3,000 m; and (4) alpine, over 3,000 m (Manual). Variation within each of these depends upon moisture, which ranges from dry, with less than 1,200 mm annual precipitation, to mesic, with 1,200-2,500 mm, to wet, with more than 2,500 mm. Thus, the combination of these major variables with local situation, e.g., soil type, can result in a wide variety of ecological niches, any one of which could provide the microclimate needed for successful establishment of a colonist.

The next in the series of Big Island volcanoes is Hualālai, whose lava flows added significantly to the western bulge of the island (see Figure 1). The mountain lies to the northeast of the town of Kailua (often referred to as Kona; in Hawaiian, *kona* refers to the lee side, in this case of the island). Visitors who have flown into Kailua International Airport have landed on lava that was extruded from Hualālai. The current peak rises to 2,521 m, which is about 400 m lower than it was at its maximum height. Thus, it was not as large as either Mauna Kea or Mauna Loa, but it did have an active existence nonetheless. Toward the end of its eruptive life, perhaps as recent as 100,000 years ago, it and Mauna Loa were active simultaneously.

Mauna Loa—the name means the long mountain—stands at 4,169 m (13,679 ft) putting it second to Mauna Kea by about a hundred meters. Despite its being second in elevation to Mauna Kea, Mauna Loa is reckoned to be the largest, single mountain mass on the Earth, as pointed out by G. A. Macdonald and D. H. Hubbard (2001) in their highly informative booklet *Volcanoes of the National Parks in Hawai'i*. Mauna Loa is one of two active volcanoes on the island, the other being Kīlauea, which we will visit next. Although most attention, both by visitors and by volcanologists, is paid to Kīlauea— undoubtedly because of easy access and spectacular activity—Mauna Loa continues to make its contribution to the island. Because of ongoing activity, the age of Mauna Loa is listed as zero to 400,000 years. Whereas eruptions are for the most part historical events on the northern volcanoes, Mauna Loa has erupted frequently in recent times. Details of eruptions dating from 1832 are listed in the Macdonald-Hubbard book with the most recent one on their list having occurred on 26 March 1984 along the northeast rift zone. A summit eruption occurred on 5 July 1975, but the impact of recent activity may best be appreciated by visiting the southwestern flank where, on 1 June 1950, an eruption began that eventually covered an area of 43.2 square miles with an estimated 493 million cubic yards of lava. The main highway crosses this area thus providing an excellent opportunity to view the extent of this eruption. The summit of Mauna Loa can be approached on foot by a trail that is described in the next chapter, or by a road that intersects the Saddle Highway in the vicinity of Pu'u Huluhulu (which we also visit in the next chapter). The road is narrow but paved (well, sort of) to about the 11,500' level where the Mauna Loa Observatory is located. The summit is another six miles on a very rough road where a four-wheel-drive vehicle is essential. There is also a hiking trail, which may provide the quicker access.

Plate 12. Steam vents at Sulfur Banks in Hawai'i Volcanoes National Park, Hawai'i.

Undoubtedly, the most visited place on the Island of Hawai'i is Kīlauea Volcano. A full array of sights awaits the visitor, including steaming vents (Plate 12), sulfur banks (Plate 13), views into the crater and fire pit (Plate 4), gaping rifts, and spectacular lava flows that are best enjoyed at night. It is also possible to hike on an excellent trail around, into, and across a volcanic crater that was formed in an eruption

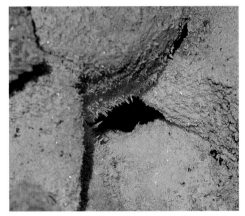
Plate 13. Close-up of crystalline sulfur at Sulfur Banks.

that began on 14 November 1959. The eruption lasted a bit over a month, during which time some 51 million cubic yards of lava and ash were vented (Macdonald and Hubbard, 1986). Thus, was Kīlauea Iki—literally, Little Kīlauea—born. Plate (14) shows a view into Kīlauea Iki from the observation deck at the end of the Devastation Trail. The scale of the depression can be measured against the group of school children on a science tour. The floor of the crater, access to which is gained either from the northern rim (near the Thurston Lava Tube) or from the southern end, is one of the best and most easily accessible places where the visitor can examine pioneer vegetation—plants that take advantage of newly created niches. Several references to native, pioneer species in and around this site are mentioned in the next chapter.

Plate 14. A view into Kīlauea Iki crater from the observation platform at the end of the Devastation Trail. Note the group of school children for scale.

The nearly constant volcanic activity of Kīlauea explains why its age is listed as zero to 400,000, as is the case with Mauna Loa. This range of ages means that it has taken about 400,000 years to accumulate the mass of material that now reaches a maximum elevation of 1,247 m, but that the volcano is still actively adding to the island. It is safe to say that it may not have reached its maximum height, although that may depend upon the continued eruptions along the East Rift Zone. To appreciate that this has been the focus of serious and recent eruptive activity one needs only note the years during which eruptive events (number in parentheses) have occurred since the Kīlauea Iki eruption: 1960, 1961, 1962, 1963 (2 events), 1965 (2 events), 1968 (2 events), 1969 (2 events), 1972, 1973 (3 events), 1974 (3 events), 1977, 1979, and 1983. The 1983 date was by no stretch of the imagination the last in this series of events! On 3 January of that year an eruption began that has continued essentially non-stop until the present. This eruption has covered over 39 square miles with a volume of lava estimated to be in excess of 2.5 billion cubic yards! (The Volcano book by Macdonald and Hubbard, from which these numbers were taken, was updated in April, 2001.) Among other results of this ongoing eruption have been the continued obliteration of the former coast road and the complete devastation of several towns perhaps the most well documented of which was Kalapana. The fate of this town was documented in the heartbreaking but beautifully illustrated book *Aloha O Kalapana* by D. Wiesel and F. Stapleton (1992). This book chronicles the efforts of local residents to save their homes, efforts that eventually proved fruitless, but which symbolized the indomitable spirit of the Hawaiian people in the face of Pele's wrath.

What does the future hold for this island? In time, movement of the Pacific Plate will take Hawai'i out of harm's way, it will cool, likely hiccup a few times, quiet down for good, grow old, and eventually be worn down to less than a mere shadow of its former self. In the mean time, however, a new island will likely have taken its place as the new kid on the block breaks the surface of the ocean. Events leading to that emergence are, of course, already underway. Swarms of earthquakes in the early 1950s centered off the southeast coast of Hawai'i signaled the possible formation of a submarine volcano. Subsequent investigations including photographic and bathymetric measurements confirmed the existence of a growing volcano. Named Lōʻihi, this volcano lies about 950 m beneath the surface of the ocean. It has already experienced formation of a caldera. What the future holds for this upstart is not clear. It is possible that it could produce a new, free-standing island, or it could coalesce with the expanding eastern coast of Hawai'i and become simply a shoulder of the present island. It will be some time, however—several thousand years in all likelihood—before Loʻihi breaks the surface. We shall simply have to be patient. (It may be of interest to the reader to learn that the name of the new volcano comes from the Hawaiian word for length, height, or distance. Considering the active forms of these words, it is easy to see that Loʻihi is, literally, a continuation of the chain of islands.)

Maui—The Valley Isle

Our next island visit is to Maui, which lies just to the northwest of Hawai'i. From the geohistorical perspective, Maui can be considered as part of a complex comprising, along with its three neighbors, Kahoʻolawe, Lānaʻi, and Molokaʻi, a much larger geological unit. At a time in the not too distant geological past—perhaps a mere half million years ago—these four were joined to form one large island, referred to as Maui Nui (literally, "greater Maui"). The association is illustrated in Figure (1) where the area within the dotted line represents an approximation of the coast line of the large island. Reference is often made to the eastern and western halves of Maui as East Maui and West Maui, respectively, almost as if they were separate places. To a degree, this is true since they were formed by eruptions of two independent volcanoes. To take this thought a bit further, it is interesting to note that the four islands comprising Maui Nui are the products of no fewer than seven independent volcanoes.

The nickname "The Valley Isle" for Maui comes from the spectacular valleys that grace both its eastern and western sides. Visitors who have spent time in the Ῑao Valley, and seen the Ῑao Needle and the shear walls that make up the valley, will have seen one of the more impressive valleys on West Maui. This valley is the inside of what is left of the caldera of a major shield volcano. The valleys on East Maui are not as immediately evident to the casual visitor, and they are not as conveniently accessible, but their formation is part of the history of another of the most-visited places on the Hawaiian Islands, Haleakalā. East Maui is totally dominated by Haleakalā, which, according to projections of size, would have been the biggest of all of the current high islands, reaching an elevation of some 5,000 m at the peak of its growth phase.

Visitors to the Island of Hawaii have the opportunity to visit the impressive volcanic crater on Kīlauea, or, for those looking for a bit more adventure, hiking into and across the smaller Kīlauea Iki crater. A good deal more challenging, but worth the effort, is a visit to the crater on Mauna Loa. By contrast, the craters of Mauna Kea and Kohala have long suffered the effects of aging characteristic of older volcanoes and no longer exist as such. This is also the case on Haleakalā where one looks—or walks— into what many guide books call the "crater," but which is technically referred to as an "erosional depression." Thus, the "crater," impressively large as it is at 11 km x 3 km and 800 m deep, is not what it appears to be. After volcanic activities in the formative stage died down, major changes in the mountain's topography were brought about by very severe erosion. Heavy rainfall led to massive stream erosion that created two very large valleys, Kaupō Valley to the south, and Keʻanae Valley to the north, as well as several somewhat lesser ones. The heads of the two larger valleys expanded and eventually joined to form the large depression. Soon after this, in terms of geological time, the mountain came to volcanic life again. This renewed activity resulted in the infilling of the depression and obscuring of the earlier features. Subsequent eruptions resulted in the many smaller, secondary cones (puʻu).

I should mention at this point that one of the "lesser" valleys referred to in the preceding paragraph, although not part of the great crater that tops Haleakalā, is impressive in its own right. Kīpahulu Valley, nearly the same size as its neighbors, drains the southeastern face of Haleakalā. The Kīpahulu Visitor Center is located at the foot of the valley and serves as trailhead for the Pīpīwai Trail along which one can visit two sets of waterfalls, Makahiku and Waimoku Falls. Visitors are not allowed beyond the second falls unless they obtain permission to use the Kīpahulu Valley Biological Preserve, which has been set aside for the study of tropical rainforests, represented in the Preserve by one of the few relatively undisturbed examples of this type of ecosystem in the islands.

One of the sights to be enjoyed on a visit to Haleakalā is sunrise over the northeastern rim of Haleakalā. The mountain is relatively free of cloud early in the morning. Later in the day, as the trade winds make their almost regular visit, clouds pour through Koʻolau Gap often bringing cold rain. The Halemauʻu Trail offers an excellent opportunity to visit Koʻolau Gap. This trail leads from a parking lot at about the 8,000' level on the mountain to Koʻolau Gap and continues into the summit depression by way of a switchback trail. This trail is mentioned in Chapter Two as an excellent track along which it is possible to see several native plants including, near the bottom of the depression, the Haleakalā silversword (*Argyroxiphium sandwicense* subsp. *macrocephalum*). Visitors should be aware of the cold, wet, and windy conditions that they might encounter on this trail (and other trails) and that hypothermia can be a real problem at this elevation, even at latitude 20°N! It is also useful to bear in mind that off-trail hiking on the Hawaiian Islands can be extremely hazardous. Broken lava, deep valleys, slippery mud and vegetation can combine to make for extremely treacherous hiking.

West Maui is the product of a somewhat older volcano, one estimated to be about 1.32 million years. Its high point at the moment, Puʻu Kukui rises to 1,764 m (5,788 ft), which is about half its estimated maximum height. Heavy rainfall on Puʻu Kukui qualifies it as one of the wettest places on Earth. Puʻu Kukui is home to several native species, many of them restricted to the small boggy area on the summit. Access to this area is difficult, involving a long, wet hike, and permission is required owing to the sensitivity of the area to habitat destruction and possible invasion by foreign species. Among the native species from this area are two species of *Argyroxiphium*, one with silver leaves, *A. caliginii*, and one with green leaves, *A. grayanum* (called greensword), and several lobelia relatives, some of which will be mentioned in Chapter Two.

Access to the Īao Valley is provided by road and the interested visitor can enjoy the view from an observation area not far from the parking lot. Access to the valleys on the northern flank of the West Maui mountains can be had via a trail along the Waiheʻe Ridge, also referred to as the "Boy Scout Trail." The Waiheʻe River drains the northeastern face of the mountain. The trail, whose plants are described in Chapter Two, follows the ridge for several miles and provides excellent views of forest, other valleys, and, if the clouds cooperate, views of some of the higher peaks. Plate (15) is a photograph of the Waiheʻe Valley at a time when the clouds were not cooperating. This is a high rainfall area and subject to strong winds. There is also a gentle trail along the Waiheʻe River that might be of interest to visitors wishing a walk without significant elevation gain.

In stark contrast to the extremely wet conditions on the eastern and northeastern slopes of West Maui, the southwestern and western coastal area lie in the rain shadow of the central massif. Here visitors find the historically important town of Lahaina, famous from its days as a whaling port and currently a major tourist destination. Farther north along the coast from the town of Kāʻanapali one finds many resorts that feature world-class golf courses. A further contrast, and one we have seen on other islands, is the comparatively gentle slope of the western flank of the mountain compared to the extremely steep and deeply cut faces on the eastern and northern flanks.

Plate 15. A view of Waiheʻe Valley, Maui from the "Boy Scout Trail."

It is not possible to separate the periods of volcanism that gave rise to the Maui volcanoes from activity that produced other islands in this group, Kahoʻolawe for example. Volcanic rocks from Kahoʻolawe have been dated at a bit over one million years (1.03 million) indicating that it was busy during roughly the same time period. Kahoʻolawe would have difficulty qualifying as a destination for vacationers, however. The island lies in the rain shadow of Haleakalā and without ground water offers a bleak landscape. It currently stands about 450 m above sea level, having reached an estimated 2,100 m when it was at its tallest. It is uninhabited, although it was occupied by a small number of people at one time, presumably when some potable water was available. Goats were introduced in the 19th century leading to the loss of much of the island's plant life, which in turn left the surface susceptible to wind erosion, which has been significant.

The island was also used by the military as a practice bombing range. After many years of protest, ownership reverted to the State of Hawaiʻi in the early 1990s following which a biological survey was undertaken. Among the botanical finds was a small legume (pea family) shrub growing on a sea stack on the southern coast of the island. Study revealed that the plant was new to science and was named *Kanaloa kahoolawensis* Lorence & Wood, recognizing its island origin (in both generic and specific epithet names!) and the two botanists involved in finding and identifying it, D. Lorence and K. Wood, both associated with the National Tropical Botanical Garden (NTBG). Only two plants were found, but fossil pollen evidence suggests that it was at one time also present on Maui and Oʻahu. Removal of goats from

Kahoʻolawe reduced one of the threats to this plant, but the unstable nature of the sea stack continues to be a threat to its survival in Nature. Seeds were collected for cultivation at the NTBG nursery.

Lānaʻi—The Private Isle

Slightly better off is the Island of Lānaʻi, which burst into life a little later than Kahoʻolawe, about 1.28 million years ago. The high point on Lānaʻi, Lānaʻihale reaches 1,027 m (3,370 ft) and is the remains of a shield volcano that at one time attained an elevation of about 2,200 m. Although blocked from rainfall to a certain extent by the high islands to the east, there is enough precipitation to support habitation. In fact, there is evidence that Lānaʻi has been inhabited since the 15th century, but the number of people that the island could support, estimated to have been about 3,000, was strictly limited by available water. Since those times, this little piece of land, only 361 km² (ca. 89,000 acres), has been the stage for a remarkable series of assaults on its ecological integrity. A recent accounting of the these events was described in 1993 by Robert Hobdy who is with the Maui District of the Hawaiian Division of Forestry and Wildlife. A brief sampling from that essay should suffice.

The first major insult occurred as the result of inter-island warfare and involved the King of Hawaiʻi, Kalaniopuʻu, and his chief, Kamehameha, who attacked Lānaʻi in 1775. The band of attackers killed most of the inhabitants, ate up or took all food and stored crops, and burned most everything else. It is useful to remind the reader that Kamehameha went on to unite the islands under one government, at the top of which he ruled as King. You may also recall his part in the Captain Cook situation described briefly in the Introduction. We will meet him again later and learn about the part he played in the sandalwood trade.

As was the custom in early seafaring days, animals were introduced to islands to provide crews with fresh meat on return visits. Provisioning of Lānaʻi began with the introduction of goats and hogs from Europe in 1778, sheep in 1791, cattle in 1793, and more goats in the early 1800s (numbers from Hobdy, 1993). European settlers began to arrive in the early 1800s as well. Near the end of the 19th century the sheep population had reached an estimated 50,000, with a large number of goats still ranging free over the entire island as well. While on the subject of animals, it is entertaining, if that is the correct word, to relate the story of a chief who lived in the uplands of north-eastern Lānaʻi whose contribution to the fertility of his people's dogs and pigs was to keep a bonfire burning perpetually on a promontory. The amount of wood required to keep this offering to the gods going for several years boggles the mind.

Returning to the 19th and early 20th centuries, it is important to note that conservation efforts were undertaken by a few committed individuals from time to time. These activities took the form of fence construction and killing off as many feral goats and sheep as possible. Some nature preserves were established, but so much harm had been done by that time that it clearly was a case of too little too late. The soil of Lāna'i is very rich and proved excellent for cultivation of pineapples (est. to have utilized 15,000 acres), the success of which venture gained Lāna'i the sometime nickname of the Pineapple Island, although islanders much prefer to have their little domain referred to as the Private Isle. Pineapples thrive in hot, dry climates and have been, as most readers appreciate, a major cash crop for the islands. Wildlife managers associated with the State of Hawai'i as well as the pineapple industry made efforts in the 1970s and 1980s to save elements of the natural forest that still existed. One of the things undertaken was the removal of goats. This had the undesirable effect of relieving competition for the Axis deer—a dozen of whose ancestors had been introduced in 1920—who then took over the task of eating the forest. Efforts to reduce the deer population have had some effect, but, as an advertisement for the natural wonders of Lāna'i gleaned from the January 1996 issue of Sky magazineinform us: "More importantly, visitors share the best of nothing—no noise or pollution, and vast empty space on an island with twice as many deer inhabiting it as people." This, presumably, makes it an "island paradise."

Moloka'i—The Friendly Isle

We can turn our attention now to Moloka'i, the northernmost of the Maui Nui complex. Moloka'i is another island produced by the activity of two large, shield volcanoes. The East Moloka'i volcano has been dated at ca. 1.76 million years and at its maximum height was 3,300 m above sea level. The current elevation of Kamakou is 1,515 m (4,970 ft) making it the highest point on Moloka'i. Stream erosion has reduced the massif to about half that height and has produced a complex array of deep valleys. The effects of wave action can be seen in the impressive sea cliffs (Plate 16) of the northern coast of East Moloka'i. These cliffs can best be viewed as one flies from

Maui to Moloka'i. The West Moloka'i volcano was much smaller, reaching only about 1,600 m elevation. It has been significantly eroded to its present elevation of about 420 m; it has been aged at 1.9 million years.

Well after the two main volcanoes had been formed, a third lesser volcano, Kaunako,

Plate 16. Steep sea cliffs on the north shore of Moloka'i.

erupted near the foot of the northern cliff face of East Moloka'i. The steep wall of the latter deflected the new lava flows toward the ocean forming the Kalaupapa Peninsula (Plate 17). This peninsula, access to which can be made by mule train down the narrow cliff-face trail, was the site of a well known leprosarium (located in the village visible in the plate). By the end of the 1870s more than a thousand people afflicted with Hansen's disease had been sent to Moloka'i (Daws, 1968).

During the Ice Age large volumes of the Earth's water were locked into massive glaciers that covered much of the Northern Hemisphere. (There is evidence that a small glacier existed on the summit of Mauna Kea, so even the Hawaiian Islands did not escape the big chill.) Normal functioning of the hydrological cycle involves evaporation of sea water with subsequent precipitation as rain and eventual return to the ocean. The formation of glaciers, however, resulted in the removal of water from the cycle, which resulted in a drop in the sea level. The effect was quite significant in the Hawaiian

Plate 17. View of Kalaupapa Peninsula on the northern coast of Moloka'i.

Islands as can be seen in Figure (1). The area within the dotted line represents the current 180 meter submarine depth. This area represents the approximate extent of land that was above water during times of low sea level. In fact, Penguin Bank was thought to have reached an elevation of 1,000 m at the high point in its life. It is clear in this illustration that all of the component islands of Maui Nui would have been open to exchange of colonizers, animal and plant, without the need to swim or be carried in the air from one island to another. In time, however, Maui Nui fragmented into the four islands that we know today. Sometime in the range of 300,000 to 400,000 years ago the large island split in two, one piece consisting of Moloka'i and Lāna'i and one piece consisting of Maui and Kaho'olawe. One to two hundred thousand years later, each of those pieces split into their constituent islands. The history of Penguin Banks is unclear, but it has been suggested that it separated (became submerged, actually) before Moloka'i and Maui became separate islands. It is also instructive to note that Maui Nui and O'ahu would have been separated by a channel only about half the width of the present separation.

Oʻahu—The Gathering Place

Oʻahu also owes its existence to a pair of shield volcanoes, one giving rise to the Koʻolau Range and one to the Waiʻanae Range. That these are ranges of mountains rather than single peaks (more or less) suggests that eruptions occurred along fault lines rather than being tightly focused around a single vent. The Koʻolau Range makes up the eastern ridge of the island and was based upon a principal volcano estimated to have begun its work about 2.6 million years ago. The steep cliffs on the windward side of the range testify to the power of erosive forces. The slope of land on the lee side of the range, the side where Honolulu is situated, is much more gentle as it slopes to the coast.

The highest elevation attained at the peak of volcanic activity in eastern Oʻahu has been estimated at 1,900 m; the present high point is 960 m above sea level. In driving around the southern end of Oʻahu toward Waimānalo and on toward Kāneʻohe, one will actually be traveling more or less at the base of what was the western wall of the ancient caldera. The eastern wall of the caldera, and almost all of its associated structure, have been eroded away by wave action and stream flow. View points at the top of the cliff (*pali*), which one reaches from the Honolulu side, offer the visitor a vista to the east over what would have been, a few million years ago, a flaming volcanic spectacle.

The range that makes up the western spine of Oʻahu is the Waiʻanae Mountains. The oldest lava underlying this range has been dated at 3.7 million years. The highest elevation of this range is 1,231 m, which represents a loss of nearly 1,000 m from the highest point. Stream erosion is evident in the many sharp valleys that characterize this range.

Kauaʻi—The Garden Isle

The next island in our survey is Kauaʻi. This is the oldest of the current high islands, having been dated at 5.1 million years. The highest point reaches an elevation of 1,598 m (5,148 ft), which is about 1,000 m less than it was at its maximum. The high point is called Kawaikini (5,243 ft), but Mt. Waiʻaleʻale receives most of the attention because of its high rainfall. Reputed to be the wettest spot on the Earth, Mt. Waiʻaleʻale receives rainfall averaging 451 inches per year (37.6 ft). Some of the water that falls on the summit accumulates in the Alakai Swamp, a boggy area home to many local native species. This spectacular area is now accessible by means of a raised wooden walkway, access to which can be had from the Pihea Trail from the north or from a combination of jeep road and trails from the west (trailhead is in the vicinity of Kokeʻe Park Headquarters). A hike in this area can be an exceptionally wet and muddy affair, but well worth the effort, as it takes the visitor into the very

heart of one of the botanically richest areas on the
islands. One of the major
drainage systems leading
away from the summit is
Waimea Canyon, called
the Grand Canyon of the
Pacific. Route 550 follows
the western rim of the
canyon from the town of
Waimea to the Koke'e
Park area. Views into the

Plate 18. View looking north at Polihale State Park, western Kaua'i.

canyon can be enjoyed at several places along this road, including from the Iliau
Loop Trail which will be described in Chapter Two when we meet the *iliau* itself, the
native silversword relative, *Wilkesia gymnoxiphium*.

After visiting Mt. Wai'ale'ale, and enjoying a sample of the daily rain storm,
it would be educational to be whisked to a spot about 25 km west. Over this distance,
and an elevation loss of perhaps 2,000 ft, one would have traveled from the wettest
place on the island to one of the driest. The fingers of land that mark the western
cliffs of Kaua'i get a mere fraction of the rain experienced on the summit. The resulting habitat is severe but nonetheless home to species of plants found nowhere else.
Limited to the westernmost extremes of two of the ridges is the only other known
species of *Wilkesia*, *W. hobdyi*.

The southwestern corner of Kaua'i is an extensive coastal plain bordered by
a magnificent beach that stretches for many kilometers south and then east along the
southern shore of the island. The western strand of Polihale Beach is noted for its
thunderous surf and the existence of its "barking sands." The visitor will have to
experience this phenomenon personally; it is not easily described! The view along
this beach toward the cliffs is illustrated in Plate (18). Two indigenous species featured in Chapter Two were photographed in the extensive dunes in this area, species
of *Ipomoea* and *Verbena* , while a third, a species of *Waltheria*, was seen in the grassy
area a little farther inland from the dunes. The road can be quite rough and muddy
at times, but four-wheel drive vehicles are not obligatory.

To the southwest of Kaua'i lies Ni'ihau, the Forbidden Isle. This small
island, off-limits to non-Hawaiians, is described as a deeply eroded remnant of a
shield volcano. Although Ni'ihau (aged at 4.89 million years) originated as a volcano independent from Kaua'i, the two are thought to have coalesced during their
maximum activities, but became separate as they subsided with age and were worn

down by erosion. Ni'ihau attained an elevation of 1,400 m at its greatest extent but is now less than 400 m at its highest point. The natural habitats of the island have been badly affected by human activity, but there is evidence from earlier collections that several native plant species existed there at one time.

The Northwestern Islands
Nihoa

Nihoa, the closest member of the northwestern extension of the island chain, lies 275 km from Kaua'i at 23°03'N, 161°55'W. Unless one were specifically looking for the wrinkles of age characteristic of a five million year-old volcanic island, the relative antiquity of Kaua'i might be overlooked, particularly in view of its spectacular tropical greenery. No one could make that mistake about Nihoa. The age of Nihoa has been reported as 7.3 million years, and it is thought to have reached an elevation of 1,300 m at its maximum growth. At a depth of about 37 m (ca. 120 ft) the underwater base of Nihoa has a diameter of about 27 km (ca. 17 miles) suggesting that it was a mountain of some substance in its day. Today its two high points—they can hardly be called peaks—rise to 256 and 279 m. The island measures roughly 1,350 m x 450 m (ca. 170 acres) and shows the ravages of erosion, with little soil and steep sea cliffs. Plant life on oceanic islands is dependent upon water, which on the high islands is provided by the trade winds most of the year and the so-called *kona* winds (from the south and southwest) during certain seasons. In the case of the low islands, however, trade winds pass over the islands because they are not tall enough to intercept the clouds. Sea squalls provide the only rainfall that these islands get. Needless to say, to survive under these conditions, plants must adapt to a desert life style for much of their lives. About 20 species of plants live on Nihoa including the palm *Pritchardia remota*, which is native to the island but which did live on other islands at earlier times. *Pritchardia* is a Pacific genus with about 25 species, 19 of which are native to the Hawaiian Islands.

Necker Island

Necker Island (23°35'N, 164°42'W) lies roughly the same distance again from Kaua'i. The age of Necker has been determined to be 11.0 million years. The Pacific has reclaimed most of the formerly 1,100 m high shield volcano. The high point on Necker today is about 84 m; the island is about as long as Nihoa but only a third as wide, which corresponds to an area of about 45 acres. The island's biota is seriously restricted with regard to plants with only five species, none native. Again, water is the limiting resource. There is evidence that both Nihoa and Necker supported human inhabitation at some time in the distant past. Dwelling and ceremonial structures have been found, and there is evidence of dry land agriculture with irrigation channels. Several authors have pointed out that potable water may well have been the limiting factor for these communities as well.

French Frigate Shoals

About 130 km farther along the chain one encounters the French Frigate Shoals. This group consists of several small coral islands, the largest of which, at 27 acres, is Tern Island at 23°56'N, 166°32'W. Other islands in this coral atoll complex include Whale-Skate Island, which has changed from two small, vegetated islands with eight species of plants in 1977 (Rauzon, 2002) into a single one that is now completely submerged. This transformation from vegetated islands to sand bar reflects the volatile nature of low islands and points out how very quickly these changes can occur. There is also a one acre volcanic remnant among this group, La Pérouse Rock (or Pinnacle), which has been dated at about 12.8 million years. It is too small to accommodate vascular plant life.

Gardner Pinnacles

Farther into the middle of the Pacific Ocean lie the Gardner Pinnacles (at ca. 25°N, 166°W). Six acres of former volcano serve to remind visitors, of which there are few other than sea birds and biologists, that this too was once a shield volcano dated at about 15.8 million years. *Portulaca lutea*, 'ihi or purslane, a salt-spray and drought-tolerant species, otherwise widespread in the islands, is the only plant species on this remnant island. Carlquist (1980) suggests that without a constant "trickle of immigrants" from other islands (new propagules) even this inhabitant might disappear.

Maro Reef

The next in line is Maro Reef which covers an area with dimensions 31 miles by 18 miles. The reef is awash at low tide and has been the site of several wrecks. Rauzon (2002) notes that the richness of marine life in the vicinity of Maro Reef is second only to that of Laysan Island to which we now jump.

Laysan Island

On the next islands we would not see any visible evidence suggesting that they were at one time high islands of volcanic origin. I specify 'visible' because investigations of the platforms upon which the next group of islands are built were necessary to reveal that they consist of volcanic rock. Laysan Island, at 25°46'N, 171°44'W, is a seabird-nesting island with an elevation of perhaps 12 m. The atoll measures nearly one and a half acres of land above high tide, but at a depth of about 200 m lies a volcanic base measuring some 210 square miles. In the days of its youth—it has been dated at 20.7 million years—Laysan was a very significant mountain.

Because Laysan has been a major seabird nesting site for millennia, huge deposits of bird excrement—guano—have accumulated. It was not long after Europeans found Laysan Island, in the early part of the 19th century, that guano miners discovered the richness of the deposits. Rauzon (2002) notes that by 1904, when the North Pacific Phosphate and Fertilizer Company was sold—for all of $1,750—more than 450,000 tons (!) of guano had been removed. Feather collectors had a similarly devastating impact upon bird life, and the importation of rabbits and Guinea Pigs, to amuse the children of the work's foreman and provide for a potential meat canning business, had pretty well taken care of the rest of the natural scene. Escaped "livestock" ate the vegetation to the ground, which, in due course, removed the source of the problem. The situation could not be better described than was done by D. R. Dickey in 1923, as quoted by Rauzon: "Verily the damned rabbits have done their worst. As far as the eye can see with the glasses and from our hurried trip down the island, there is **not a living bush or twig or spear of grass left** on the whole island outside of the two poor palm trees and 3 bushes near the house…. In my wildest pessimism I had not feared such utter extirpation of every living plant." [Emphasis in the original.]

Despite the difficulty of returning an island that had suffered such abuse to some semblance of its earlier self, efforts toward that goal have had some success. One of the projects, described in the first person by Mark Rauzon, was the all-out war waged on the weedy grass called the common sandbur, *Cenchrus echinatus*. This species is widespread on all of the main islands as well as essentially anywhere on the outer islands where disturbed sites exist. The species is draught tolerant and has a well developed root system (for which reason it has found use as a soil stabilizer elsewhere). A hands-on weeding program was successful in eradicating sandbur from the island. Without constant vigilance, however, reintroduction could occur in view of the widespread occurrence of the plant on other islands and the apparent ease of movement. Other species that were present on the island at the time of the first European visits included the Nihoa palm *loulu* (*Pritchardia remota*), a species of sandalwood (*Santalum ellipticum*), as well as an array of beach species common on Pacific islands.

Lisianski Island

About 185 km to the northwest of Laysan lies Lisianski Island (26°04'N, 173°58'W). Lisianski has been dated at 23.4 million years. This island's recent history is similar to that of Laysan, except that it hasn't experienced the extreme ravages of the latter. It has had its share of visitors looking for feathers, a few wrecks, an episode of rabbits that ended when edible greenery ran out, and an explosion of house mice. I include no quotation for the mouse story; it must be read to be appreciated fully (Rauzon)! Lisianski Island is also extremely rich in bird life and thus a potential source of guano. The magnitude of the deposit on Laysan, however, diverted attention from Lisianski; as well there were problems getting a license to mine the smaller island. Thus, Lisianski was spared the devastation that turned Laysan into a wasteland.

Pearl and Hermes Reef

Pearl and Hermes Reef (27°48'N, 175°51'W) is a true atoll with a few small sand islands, a fringing reef, and a lagoon. The volcanic base upon which the atoll rests has been estimated to be about 20 miles long and 12 miles wide which indicates that a substantial mountain existed at one time. Southeast Island is the largest of the islets with an area of 34 acres. This area has been a source of oyster shells, which were of value for the manufacture of buttons. The land area was too small to accommodate guano mining although there is a significant resident seabird population.

Midway

Midway Atoll is likely the only member of the northwestern chain that would be known to most people owing to the part it played in World War II. In June 1942 Japanese and American aircraft carrier-based planes fought a fierce battle that resulted in heavy losses to both sides but significantly greater losses in terms of carriers to the Japanese. Defense of Midway prevented an inevitable invasion of the main Hawaiian Islands. The history of Midway as an important and strategic mid-oceanic stopping point began many years earlier, however, when a refueling station was established for transoceanic shipping, and facilities were built to accommodate military aircraft as well as commercial aircraft flying across the Pacific Ocean. Midway also served as a trans-oceanic cable station. Although Midway is geologically one of the Hawaiian Islands, it is not part of the State of Hawai'i; rather it "belongs" to the United States government.

Midway Atoll is located at 28°12'N, 177°24'W, and, as its name implies, it lies roughly midway between North America and Asia. The atoll, first visited by Europeans in 1859, consists of two large islands, Sand and Eastern Islands, and one very small one, Spit Island. Although it was widely believed for many years that the low islands of the Hawaiian chain are the remnants of volcanoes, direct proof was not obtained until the 1960s when rock cores obtained by drilling at sites on Midway revealed basaltic rocks (of volcanic origin) underlying the coral. At one of the drilling sites, lignitic clays that had been deposited under swampy conditions were observed to overlie the lava. Midway has been dated at about 28.7 million years.

Original vegetation would have been typical of coral atoll vegetation of the Pacific Basin, i.e., based upon plants that have the capacity for long distance dispersal by birds and by ocean drift. Modification of the flora of Midway has been extensive, starting with attempts to make it a more attractive place to live for administrative staff and for people associated with the shipping and air stations. Ironwood trees (*Casuarina equisetifolia*) were planted to serve as windbreaks and topsoil was imported for gardens. At the present time there are about 200 species of plants on Midway, 75% of which are weeds, ornamentals, or vegetables (see Rauzon for details).

Kure Atoll
Kure Atoll (28°35'N, 178°10'W) lies to the northwest of Midway. It is the next oldest of the low islands dated at 29.8 million years. Kure consists of two main islands, Green and Sand, plus a few much smaller ones. Green Island is large enough to have accommodated an airfield and served for some time as a Loran antenna station. The atoll's main claim to fame would seem to be the number of ship hulls it has accumulated over the years, but the first "official" disaster was recorded in 1837. Plants on the islands are typical of oceanic island floras with a significant growth of beach *naupaka* (*Scaevola sericea*).

A National Wildlife Refuge

In recognition of the incredible richness and variety of bird life on the northwestern islands, President Theodore (Teddy) Roosevelt declared them the Hawaiian Islands Bird Reservation in 1909. In 1940 President Franklin Delano Roosevelt changed the name to the Hawaiian Islands National Wildlife Refuge, which had the effect of including sea turtles, monk seals, and other wildlife under protection. The refuge is managed by the United States Fish and Wildlife Service (USFWS) and is off-limits to both commercial and pleasure craft. Strict control of access is maintained which, with the help of numerous volunteers and members of the USFWS, has greatly helped in returning the islands to pre-exploitation level.

An Historical Overview

How far can we carry this on before we run out of islands, or come to a gap in the island building process itself? Indeed, there appears to have been at least one gap. Going beyond Kure Atoll, the next high island seems to have been the Koko volcano, one of the members of the southern Emperor Chain. The age of the Koko seamount has been determined to be about 49 million years, which means that there was a gap of about 20 million years during which island building appears to have been on hold. By the time the hot spot resumed its activity, any high islands on the conveyor belt would have been moved far to the northwest and been degraded to the status of atoll. After the hot spot returned to more or less full activity, there still were gaps in the production of high islands. According to the recent analysis of the archipelago's geological history by Jonathan Price from the University of California at Davis and D. A. Clague from the Monterey Bay Aquarium Research Institute (2002), there were few islands that exceeded 1,000 m in height between 32 and 18 million years ago. Gardner Pinnacles is the remnant of an island that existed about 16 million years ago. It has been estimated that at its peak it exceeded 4,000 m elevation and was about the size of Hawai'i, and was the largest of the high islands before the current batch. Between 18 and 8 million years ago there were several peaks that exceeded 1,000 m and may have approached 2,000 m in elevation. Necker Island, about 11 million years ago, was a member of this group of medium sized islands. There were no peaks of significance between Necker's days of glory and the emergence of Kaua'i. Another factor discussed by Price and Clague is the spacing of volcanoes (islands). There were only three instances when more than three volcanoes were connected above the surface: (1) St. Rogatien-Perouse (St. Rogatien Banks lies half way between the French Frigate Shoals and Gardner Pinnacles); (2) O'ahu-Maui; and (3) Hawai'i. The proximity of volcanoes is, of course, important for consideration of inter-island colonizations. The first of these occurred during an earlier phase of island building, while the latter two occurred during the more recent peak of volcanic activity.

The fact that high sites above 1,500 m elevation had likely vanished before Kaua'i formed suggests that the present flora and fauna of the islands, for the most part, must be comparatively recent in geological terms. Price and Clague compiled a list of all groups of organisms for which reliable age data exist, and from these lists it is easy to see that the majority of the groups have colonized the islands since Kaua'i has existed, i.e., since about 5 million years ago. Exceptions are the Hawaiian fruit flies (26 million), the lobelia group (15 million), and the damselflies (9.6 million). Values for the mints, 2.6-7.4 million, represent the range of ages taking experimental error into consideration with the average value within the range of "recent" arrivals. The silversword alliance clocks in at 5.1 million, and from there on values get smaller until one reaches *Metrosideros* ('ōhi'a) at about a half million years. It is interesting to see that no bird lineages need go any further back than Kaua'i. Twelve of 15 multi-species lineages of birds have diverged within the lifetime of Kaua'i; the others have all colonized since then. The most important lesson to take away from all this is that we are dealing with a biota that is comparatively new, with a few notable exceptions. What future research will reveal about the ages of other island groups remains to be seen, but a vision of the islands as sites of active evolutionary change over a relatively brief period of geological time is clear.

The reader will also certainly appreciate that we are dealing with an exceptionally dynamic system—a system comprised of land masses that appear, provide homes for a variety of organisms, grow old, and disappear, only to be replaced by the next island, and then the next. We know the ages of the individual islands, or volcanoes from which they sprang, and we know where the various islands lay with regard to each other. We have also seen that they have not always existed as isolated entities, but may have been variously attached to neighbors, or at least were closer together at one time than they now are. Another factor to bear in mind is that elevation is also a dynamic feature of an island's life, as the repeated reference above to present and historical elevations of the founding volcanoes shows. For example, although neither Kaua'i nor O'ahu has alpine habitats at the present time, they would have had them when they were younger. Over time, however, erosion and subsidence would have eliminated those habitats. Survival of species inhabiting those sites would have required adaptation to conditions at lower elevations, or migration to a younger island where suitable alpine habitats were available. Both processes have occurred. There is evidence that some alpine colonizers, *Tetramolopium* in the sunflower family for example, have differentiated to species that now inhabit lower elevation sites. Studies of other species will likely show a common thread. There is also extensive documentation that colonization of younger islands from older islands in the chain is a common theme in the evolution of the Hawaiian flora. An example of this can be seen in the silversword alliance. This group has been represented on the islands for at least five million years, yet several species of *Dubautia*, as well as both subspecies of the silversword itself, *Argyroxiphium sandwicense*, are known only on the youngest islands, Hawai'i and East Maui.

This would be a perfect place to finish off the introduction and move directly into a description of the biota of an undisturbed island system. But that is not how it turned out. Trouble was on its way. Unfortunately, some of the next colonists to arrive came in boats, carrying with them the wherewithal to establish a new homeland. Land was cleared for cultivating the crop plants they brought with them, and the destruction of the flora of the Hawaiian Islands began. A few centuries later more visitors came, some of whom settled, but others, recognizing the commercial possibilities of monk seals and sandalwood, to site only two examples, began harvesting these natural treasures in earnest. To this day, the biota of the islands continues to be subjected to threats from an expanding population, agriculture, and commercial development—no small amount of which is for the tourist industry. Not everything has been destroyed, of course, but many of the plants present at the time of the first human visitors are under severe threat. Fortunately, some are holding their own. In Chapter Two we will look at a selection of native species that have survived and others that have managed to survive but are at serious risk.

Native Hawaiian Plants

In this chapter we will meet a selection of native Hawaiian plants—by no means all of them—but enough I hope to provide the reader with an idea of the botanical riches that make these few small islands their home. The examples chosen represent a fair cross section of the sorts of plants you might be likely to find on a trip to the islands, without the need to hire a personal guide (or a helicopter) or put yourself in harm's way. Most of the sites described are easy to get to by automobile—Kilauea Iki and Thurston Lava Tube area and the Mauna Loa Trail on Hawai'i, Haleakalā on East Maui and the "Boy Scout Trail" on West Maui, and the Awa'awapuhi, Pihea, and Alakai Swamp Trails on Kaua'i (although the latter can be tricky in wet weather). A visit to the National Tropical Botanical Garden and Limahuli Garden, both on Kaua'i, will also introduce the visitor to a variety of native species.

The first part of the chapter will focus on native species; the second part will feature species that occur elsewhere in the Pacific Basin but are also prominent members of the Hawaiian flora. In addition to learning something about the biology of the plants, we will speculate on where they may have come from, who their ancestors might have been, and when those ancestors may have arrived on the islands. Some recent studies focusing on these questions will be discussed. We will also see how the Hawaiians used some of the plants that grow naturally on the islands, as well as a few that they brought with them. We will also look at how some of the plants got their names, whether scientific or Hawaiian.

It is useful to start this chapter with an overall description of the flora of the islands. Three kinds of plants have to be considered: those that are native (syn: endemic) to the islands, those that are indigenous to the islands, by which we mean plants that occur elsewhere in the Pacific Basin but got to the Hawaiian Islands by natural means, and those that occur on the islands but had human help in getting there. The latter category includes plants brought to the islands by the original Polynesian settlers as well as those brought in either by accident, or by people hoping to establish plantations with financial gain in mind. Many of these plants have become very serious pests. The term "naturalized" refers to alien species that have taken well to their new home—often with enthusiasm—and reproduce new generations readily.

According to the *Manual*, the flora of the Hawaiian Islands consists of 1,817 species, of which 956, or about 53%, are native (endemic and indigenous). The remaining 861 are naturalized species, some of whom we will encounter in Chapter Three. Referring again to the *Manual*, we see that of the 956 native species, 850, or 89%, occur only in the Hawaiian Islands. Some workers have the percentage a bit higher, but even the most conservative number ranks as the highest level of insular species endemism on the planet! Native species represent 216 genera of which 32, or 15%, occur only on the islands. In simple words, this means that these groups of organisms have diverged so much from their ancestors—either continental or from other island groups—that they have been recognized as unique assemblages at the level of genus. The species comprising these genera are, of course, also natives. I should add that there are about a half dozen genera that have all but one species on the Hawaiian Islands, with the remaining member native to some other island in the Pacific Basin, e.g., Tahiti. The level of generic endemism is also the highest for any island system in the world.

At the next higher level, the *Manual* includes information on 146 plant families, all of which are known elsewhere in the world. Plant families native to islands are very scarce items, perhaps as many as a dozen could be argued. This scarcity can be explained quite simply: in order to qualify for recognition as a family, a group of plants must have accumulated a significant number of unique features. There simply hasn't been enough time for plants on the Hawaiian Islands to have diverged (evolved) enough for family recognition. A comment on the absence of native families on the Hawaiian Islands might be of interest to the reader unfamiliar with the "name game." There is nothing to stop someone from suggesting that some group of plants on the islands has diverged enough for family recognition. However, the suggestion would have to be accepted by the broader botanical community, which is not likely, given the conservative nature of that august body. You will note in a few instances below that suggestions have been made for the recognition of certain genera as being uniquely Hawaiian, e.g., *Neurophyllodes* for the six native species of *Geranium*, and *Lysimachiopsis* for the Hawaiian native species of *Lysimachia*. Very convincing evidence, indeed, must be presented before such proposals become widely accepted. Changes in the other direction have been suggested as well, as in the case of the Hawaiian native genus *Pelea*, whose many species are now considered to be members of the widespread genus *Melicope*. In that instance sufficient evidence was accumulated to support the change. The name *Pelea* is still maintained, however, but only as it refers to the native Hawaiian species of the larger genus which are now lumped together as subgenus *Pelea*. Thus, the subgenus of that name consists solely of native species.

The five most common families represented on the Hawaiian Islands, with the number of native species indicated, are the sunflower family (92), the grasses (47), campanula relatives (110), the legumes, or pea family (20), and the mints (54). These families are very common components of continental floras, which makes sorting out ancestral relationships even more challenging.

An interesting question that has led to some speculation deals with how many original colonization events were needed to arrive at the current native flora. Authors of the *Manual* suggest that the answer lies in the range 270 to 280 colonists, although I have seen an estimate as high as 291. Even the higher estimate seems, at first glance, to be too low to account for the variation that one sees in the native flora. What this number forces us to consider, however, is the extraordinary extent of the evolutionary changes that have occurred since those relatively few original colonists arrived on the islands.

The obvious next question is, where did the colonizers come from? One might think that North America would be a likely source for many of the island natives, owing to its comparative proximity, but F. R. Fosberg (1948), former curator of Pacific botany at the Smithsonian Institution, estimated that only about 18% of Hawaiian plants have obvious affinities with North American species. Far greater affinities are thought to lie with groups native to the southwestern Pacific Ocean (Malesia), with others being related to species from Australia, New Zealand, southern South America, and southeastern Asian tropics. As well, there are a few genera whose affinities are so obscure that one can only guess at who the ancestors might have been.

A problem that confronts us in studying the Hawaiian flora is that quite a number of native species are extinct; what we know about them is restricted to collectors' records and dried specimens deposited in herbaria. In some instances it appears that the original collections—likely done in an enthusiastic moment of discovery—were the cause of extinction! More on this idea in the discussion of Hawaiian mints below. We will start our survey of native Hawaiian species with the silversword alliance, one of the most thoroughly studied groups in the Hawaiian flora, and certainly one of the most unusual.

The silverswords

The silversword alliance, described in detail by Prof. Gerald Carr of the University of Hawai'i (1985), consists of three genera, *Argyroxiphium*, *Wilkesia*, and *Dubautia*. *Argyroxiphium*, the silverswords and greenswords, consists of four or five extant (living) species, *Wilkesia* consists of only two species, while the largest of the three, *Dubautia*, consists of at least 23 species. We will look at these genera in that order with comments on diversity of growth form and habitats and descriptions of places where some of them can be easily seen in the wild. Following the descriptions will be a discussion of the evolutionary history of the group, including recent research using DNA data that firmly establishes that their ancestors immigrated from California.

Argyroxiphium-āhinahina

The Hawaiian name for *Argyroxiphium*, *āhinahina*, is based on the simpler form *āhina* meaning gray, gray- or white-haired (*P. & E.*). There was no word in the Hawaiian language in earlier times for silver, so the term for gray was used to describe the plant's appearance. The silvery leaves of this species are illustrated in Plate (19) where a small colony has been photographed along with a colleague to show scale. The term *āhinahina* is also used to describe other plants that have similar appearances, such as *Artemisia mauiense*, which also has white woolly leaves. (Note: *Artemisia* in North America is the common "sage brush" of the dry western prairies.)

Plate 19. *Argyroxiphium sandwicense* ssp. *sandwicense*, the Mauna Kea silversword. This picture shows the typical clustering of young rosettes.

The botanical name of the genus, *Argyroxiphium*, consists of two parts from the Greek, "argyros" = silver and "xiphium" = dagger, in reference to the long, dagger- or sword-like leaves covered in fine, silky hairs which give them a silvery appearance. The hairs serve an important function in protecting the leaves from the full impact of bright sunlight at the elevations where this species lives, in the neighborhood of 9,000 to 10,000 feet on

Plate 20. Early flowering head development in *A. sandwicense* ssp. *sandwicense*.

Plate 21. A later state in flowering head development in *A. sandwicense* ssp. *sandwicense*.

Mauna Kea on Hawai'i and Haleakalā on East Maui. They also serve as protection against desiccation, a serious problem at higher elevations where, in addition to strong sunlight, a plant is frequently subjected to strong winds. The specific epithet (some people call this the "species name") *sandwicense* reflects the geographic origin of the species, the Sandwich Islands. Three species of *Argyroxiphium* exhibit silvery hairs: *A. sandwicense*, who we have just met, *A. caliginis*, the 'Eke silversword from the summit bogs of West Maui ('Eke is the name of one of the craters on the summit of West Maui), and *A. kauense*, the Ka'u silversword from a few sites on the flanks of Mauna Loa (the southernmost volcano on Hawai'i). The two remaining species, *A. grayanum*, from higher elevations on West Maui, and *A. virescens*, now possibly extinct, but formerly a resident of East Maui, are called "greenswords" to indicate that their leaves, although sword-like, lack the fine hairs that characterize the other three species.

    Argyroxiphium belongs to a group of plants whose life style is described as "monocarpic." This term refers to a plant that grows for a number of years, perhaps 20 or more as in the case of *A. sandwicense*, eventually develops a large flowering head, sheds its mature seeds, and then dies. Plate (20) illustrates an early stage in the development of a flowering head. Plate (21) illustrates a somewhat more advanced stage of development where the individual composite flower heads are visible. A rosette of silvery leaves and large, mature flowering head are illustrated in Plate (22);

Plate 22. A fully developed flowering head of *A. sandwicense* ssp. *sandwicense.*

a close-up of its composite flower head appears in Plate (23). This is the only genus of the alliance that possesses ray florets (the petal-like, pigmented structures in the photograph). *Wilkesia gymnoxiphium*, to be met a little later, enjoys a similar life style, except that its rosette of leaves appears at the top of a long stalk. It is also a monocarpic species. Some other species of *Argyroxiphium* and the only other species of *Wilkesia* are polycarpic, which means that they do not die after producing a crop of seeds.

Plate 23. Close-up of a single composite flower of *A. sandwicense* ssp. *sandwicense*.

A few comments might help explain how some of these plants got their names. The derivation of "silversword" was mentioned above, but it is necessary to go a little further in order to explain the full name of the species, *Argyroxiphium sandwicense*. This expanded name tells us that this particular plant occurs in the Sandwich Islands, the name originally given to the islands by Captain Cook to honor his patron, the Earl of Sandwich. The term *sandwicense* is the Latinized form. The plant was first collected high on the flanks of Mauna Kea in 1825 by James Macrae, botanist on the H.M.S. *Blonde*, who recorded in his journal that it is "…truly superb, and almost worth the journey of coming here to see it on purpose." This specimen made its way to the continent where the French botanist de Candolle, in 1836, gave it the name by which it is now known. Specimens were also collected on Mauna Kea in 1834 by David Douglas; these specimens were eventually delivered to Dr. W. J. Hooker, a world's authority on plants of the Southern Hemisphere and Islands of the Pacific, whose home base was the Royal Botanic Gardens at Kew in England. Not knowing of the name already given the plant by his French colleague, Hooker, in 1837, called the new species *Argyrophyton douglasii*, which was his way of recognizing Douglas' contribution by naming the new species after him (a common form of recognition in the botanical business; more on this practice below). Literally, the Latin name means Douglas' silver plant, "argyros" as above and "phyton" meaning plant. Learning of the earlier name, however, Hooker withdrew his suggestion. In 1861 Asa Gray, a botanist at Harvard University, described *A. macrocephalum*, which were specimens of the Haleakalā silversword collected by botanists of the United States Exploring Expedition of 1838-1842, recognizing that this plant has a large flower head (macro = large; cephalum = head). It is first-come first-served in the plant naming business, however, so the botanical community was stuck with de

Candolle's name. But we are not finished with "*macrocephalum*" as a useful term. Recently, differences between plants growing on Mauna Kea (Hawai'i) and those growing on Haleakalā (East Maui) led to the recognition of two subspecies, subsp. *sandwicense* to recognize plants from the original sites, and subsp. *macrocephalum* from Haleakalā. In addition to having a large head, just mentioned, the heads are also differently proportioned. The full name for this subspecies, which recognizes Asa Gray's first use of the term *macrocephalum*, and the contribution by the Hawaiian botanist A. K. Meyrat and colleagues at the University of Hawai'i (Meyrat et al., 1984) that size differences were significant, is *Argyroxiphium sandwicense* DC. subsp. *macrocephalum* (A. Gray) Meyrat.

As has been the case with many other Hawaiian native plants, *Argyroxiphium* species have not had easy lives, at least not since the arrival of Europeans. Remember that one of the five known species is possibly extinct. Consider that the others might well be in the same category were it not for extraordinary conservation efforts. Owing to the pressures of cattle grazing and the disruptive behavior of feral pigs and goats, habitat loss has been severe. The plant's enemies were not restricted to feral animals by any means! There was a time when revelers found pleasure in rolling the plants down the slopes of Haleakalā (Kepler, 1998). At one time *A. sandwicense* grew in extensive colonies on Mauna Kea between about 8,000 and 10,000' elevation. Today, the situation is drastically different; a good deal of effort is required to see silverswords on Hawai'i, where they survive today in fenced exclosures on the flank of Mauna Kea at an elevation of about 9,000'. Access to these sites requires travel on a very rough jeep road not recommended for the casual visitor (rental car companies frown on this kind of adventure as well). The sites consist of a few acres of ground within which perhaps a few hundred *A. sandwicense* plants live in a goat- and pig-free environment. We didn't see any animals on our visits to these sites, but they had been there, as was obvious from the badly disturbed ground outside the fence, nibbled shrubs, and bits of goat and pig hair caught on a few of the trees that grow there and on the exclosure fence itself.

An even more extensive project to protect silverswords and other native species involved erecting many miles of fencing around the summit of Haleakalā on East Maui. The perimeter is patrolled periodically in order to find and repair breakages in the fence where animals could get through and venture into the sensitive areas. A visit to Haleakalā is the easiest way to see silverswords—thousands can be found involving no more than a two to three hour hike from the Visitor's Center on the Shifting Sands Trail. A few are usually on display in a garden at the Park Headquarters and in a small garden at the Summit Observation area. The plants are also featured on the Silversword Loop which involves a somewhat longer hike.

Plate 24. An out-planted *A. sandwicense* ssp. *sandwicense* on Mauna Kea.

Plate 25. A field of out-planted *A. sandwicense* ssp. *sandwicense* on Mauna Kea.

In a cooperative effort, private and government organizations have undertaken an extensive out-planting exercise involving both the Mauna Kea silversword (*A. sandwicensis*) and the Mauna Loa silversword, *A. kauense*. (Robichaux et al., 2001; Purugganan et al., 2002). Following hand pollination of plants in the field and in cultivation, seeds were collected, allowed to germinate in cultivation, and the seedlings set out in Nature in sites originally occupied by these species. Over 4,000 Mauna Kea silversword seedlings and over 11,500 Mauna Loa silverswords have been outplanted according to the latter reference. It is heartening to learn that establishment has exceeded 90% in some of the sites. Plate (24) shows a flagged plant that is part of an out-planting on Mauna Kea. Plate (25) shows a field of red flags, each indicating a recently planted seedling at the same site. Additional information on the silversword program can be found at http://pacificislands.fws.gov/wesa/mkeaslvrswrdidx.html.

Dubautia-na'ena'e, kūpaoa

The *Manual* lists 21 species of *Dubautia*, (23 according to a more recent review, Carlquist et al., 2003) which, with the recognition of several subspecies, makes this the largest genus of the alliance. It is also one of the larger genera of native species in the islands. There may in fact be others, but additional study of remote areas are necessary. There appear to be no extinct species in this genus, but several are considered rare and/or endangered and likely need special attention to keep them on this side of extinction. One of the rarest species is *D. latifolia* (latifolia = broad leafed), an extremely unusual species owing to its being a liana. The climbing growth form is not common in the sunflower family and is seen in no other member of the silversword alliance. The *Manual* informs us that this species is in danger of being extirpated by one of the most aggressive of the alien invaders, the rapidly growing climber commonly called banana poka. Banana poka, a member of the passion fruit family, will be discussed in Chapter Three.

Hawaiians referred to all members of *Dubautia* as *na'ena'e*, referring to the aroma of the flowers, or *kūpaoa*, which refers in general to certain strong-smelling plants including *Dubautia*. When wishing to be more specific, the Hawaiians added descriptive terms that referred to color, e.g., *na'ena'e pua melemele* for yellow-flowered *Dubautia* (*D. laxa*), and *na'ena'e pua kea* for white-flowered *Dubautia* (*D. paleata*). (Note that we have seen "*kea*" before in Mauna Kea, the white mountain, referring to snow cover; and *pua* is the Hawaiian word for flower.)

Some species of *Dubautia* are very easy to find in the wild; in fact, if you look in the right place, *D. scabra* may be one of the few plants visible. This species can be readily seen on Hawai'i as you walk along the Devastation Trail from the parking lot toward the Kīlauea Iki observation platform. Kīlauea Iki, who we met in Chapter One, came into life violently in November of 1959 providing the hill of cinders that lies directly ahead of you. In addition to creating the new habitat, the eruption devastated whatever grew in the vicinity previously. *Dubautia scabra* belongs to a group of plants known as pioneer species—plants that are among the first to colonize newly established habitats. *Dubautia scabra* is a low, shrubby plant with a mass of small white flower heads and often a lot of old, dead plant material around the base (Plate 26). Another plant that you are

Plate 26. *Dubautia scabra* near the Devastation Trail boardwalk, Hawai'i Volcanoes National park.

Plate 27. *Vaccinium reticulatum*, *'ōhelo*, showing the bright red berries on plants growing near the Devastation Trail, Hawai'i Volcanoes National Park.

Plate 28. A fern (*Nephrolepis* sp.) living in a tree hole formed when hot cinders burned out the tree. Devastation Trail, Hawai'i Volcanoes National Park.

Plate 29. A hybrid between *Dubautia scabra*, which has white flowers, and *D. ciliolata*, which has dark yellow flowers. This plant was photographed near the end of the Devastation Trail.

likely to see growing in this area is a species of blueberry (*Vaccinium*, 'ōhelo), recognizable by its often brilliantly red-colored terminal leaf clusters and equally red berries (Plate 27). You might also come upon ferns growing in tree holes, pits made when trees were burned out by the hot cinders deposited by the volcanic eruption (Plate 28). With ample moisture, rich soil, and the protection provided by the tree well, the young fern (a species of *Nephrolepis*) has an excellent chance to become established. A little farther along the boardwalk, as you enter the forest again, you may see another plant that resembles *D. scabra*, except that it is somewhat taller and its flowers are pale yellow (Plate 29). This is likely a hybrid between *D. scabra* and *D. ciliolata*. *Dubautia ciliolata*, a shrub characterized by yellow flowers and a more upright posture, grows on lava fields that are older than those you crossed on the Devastation Trail boardwalk. (*Dubautia scabra* and *D. ciliolata* each consist of two subspecies, but only one of each occurs in the area that I am describing here.) *Dubautia ciliolata* can easily be found growing in the vicinity of the Hawai'i Volcano Observatory/Jaggar Museum (a must visit!), which is situated on the western rim of Kilauea Crater. In addition to excellent specimens in the planted area at the Observatory (Plate 30), a short stroll across the road will provide a look at plants growing in more typical, wind-swept habitats where they occur with two other natives, 'ohi'a lehua (*Metrosideros*) and more plants of 'ohelo (*Vaccinium*). More on both of these plants later in this chapter.

Plate 30. *Dubautia ciliolata*, growing in the planted area at the parking lot of the Hawai'i Volcanoes Observation Center/Jaggar Museum.

Plate 31. Three pioneer species in a crack on the crater floor of Kilauea Iki: the fern *Nephrolepis, Dubautia scabra,* and *Vaccinium reticulatum.*

As we saw with the fern living in a burned-out tree hole above, plants take advantage of whatever circumstances exist. Other excellent examples of pioneer plants becoming established can be easily seen on the floor of Kīlauea Iki. As the lava lake within the crater cooled, its surface underwent buckling resulting in many cracks and crevices. Seeds or spores finding their way into these cracks and crevices found themselves in well protected niches ideal for colonization. Plate (31) illustrates this phenomenon with the familiar fern *Nephrolepis* in the foreground, next to which is *Dubautia scabra,* and last in this little garden is *Vaccinium reticulatum.* Over time these plants will provide litter into which other species' seeds or spores may fall and find a home. On first glance the floor of this and other young craters appear black and sterile, as the picture of Kīlauea Iki taken from the rim might suggest (Plate 14), but as we have just seen, closer examination reveals a far different picture. In Chapter Three we will meet several alien species—one of which, *Buddleia asiatica* (Plate 182), was photographed within the crater.

The two *Dubautia* species just met occur on lava fields of slightly different age, *D. scabra* on the newer lava, *D. ciliolata* on somewhat older lava. This reveals the sensitivity of obviously similar species to subtly different ecological requirements and/or tolerances. This barely scratches the surface of habitat diversity for the genus, however. The two species just mentioned grow in semi-desert conditions, while at

Plate 32. *Dubautia raillardioides* in the rain forest along Pihea Trail on Kaua'i.

the other extreme is *D. waialealae*, which grows in the summit bog of Mount Wai'ale'ale on Kaua'i (ca. 5,200' elev.) considered to be the wettest place on earth with over 460 inches of rainfall per year (some sources say 600 inches). In addition to the widely different ecological situations in which representatives of this genus can be found, growth forms equally vary. In addition to bog plants, there is a small tree (*D. arborea* appropriately), the semi-desert species just described with their short, sharp-tipped leaves, a rain forest species with long grass-like leaves, and the liana *D. latifolia*, to mention only the extreme cases. *Dubautia scabra* and *D. ciliolata* are easy to find in the vicinity of Kīlauea Volcano as described above.

Dubautia menziesii, which resembles *D. scabra* and *D. ciliolata* in having sharp-tipped leaves arranged in regular whorls, grows on the upper slopes of Haleakalā and within the "crater" itself. (Note: The so-called "crater" is actually an erosional

Plate 33. A view from the Kalalau Lookout on the Pihea Trail on northern Kaua'i showing the rugged terrain characteristic of old, heavily eroded volcanic mountains.

NATIVE HAWAIIAN PLANTS 41

depression. See Chapter One.) It can also be found along the Halemau'u Trail on Haleakalā, mentioned above as one of the ways into the crater. An easy walk along this trail (and return to the parking lot) provides the visitor with an excellent opportunity to see *D. menziesii*, along with other native species—one of which, a species of *Geranium*, will be met below. To see a species of *Dubautia* that has a totally different growth form, *D. raillardioides* (Plate 32), one must invest a bit more effort. This species, which grows on the margins of bogs on Kaua'i, can be found at about the one mile marker along the Pihea Trail. This trail starts at the end of Route 550 and skirts the edge of the cliffs that define the northern face of Kaua'i. Along this trail one passes the viewpoint into the Kalalau Valley (Plate 33), likely one of the most photogenic sites in the Hawaiian Islands. The ruggedness of this terrain makes field studies very difficult and extremely dangerous.

Wilkesia-iliau

Wilkesia, named in honor of Captain Charles Wilkes, commander of the United States Exploring Expedition of 1838-1842, consists of two species, one common and easy to find, and one not so easy to find. Both species are native to dry western parts of Kaua'i. *Wilkesia gymnoxiphium*, whose Hawaiian name is *iliau*, resembles the silverswords in having a thatch of leaves out of the top from which emerges a flowering head. In the case of the *iliau*, however, the tuft of leaves sits on top of a stalk, sometimes as much as a meter and a half long (Plate 34). Plate (35) shows a plant before the flowering head has fully opened. It is not uncommon to see plants at var-

Plate 34. *Wilkesia gymnoxiphium* in full bloom at the Iliau Loop Trail on Kaua'i.

Plate 35. *Wilkesia gymnoxiphium* in bud at the Iliau Loop Trail on Kaua'i.

Plate 36. A plant of *Wilkesia gymnoxiphium* with the Waimea Canyon as background.

ious stages of opening. Occasionally, plants can be seen whose stems have branched to produce a candelabra-like effect. The floral structure emerges from the top of the stalk in much the same way that we have seen in the silversword. *Iliau* is one of the easiest of the Hawaiian native species to locate in the wild since it is the featured plant along the Iliau Loop Trail on Kaua'i. This small park can be found 6.3 miles south of Koke'e Park Headquarters. The Iliau Loop Trail also serves as the trailhead for the Kukui Trail, which provides hiking access to Waimea Canyon (Plate 36). There is also a small population of *iliau* on the high ground at the Y-junction of Rts. 550 and 552, but the population is not as impressive as the one at the Loop Trail, and there is no trail as such. A population can also be enjoyed near the end of the spectacular—and not to be missed—Awa'awapuhi Trail, the entrance to which is 1.5 miles south of the Koke'e Park Headquarters.

The second member of this genus, *W. hobdyi*, was described as a new species only in 1971 based on specimens collected by Robert Hobdy on Polihale Ridge, one of the fingers of land that reach toward the Pacific Ocean on the extreme western cliffs of Kaua'i (Plate 37). By contrast, *iliau* was first collected in 1840 by members of the United States Exploring Expedition and named in 1852 by Asa Gray at Harvard University. In addition to being a lot more abundant, collection of *iliau* does not pose any immediate threat to life and limb, not something that can be said about the *W. hobdyi* sites! According to the *Manual*, *W. hobdyi* is thought to consist of only about 250 plants. In my visits, only a dozen or so plants were seen, although we did not perform a census. The presence of feral goats was much in evidence, however. It is not

Plate 37. *Wilkesia hobdyi* grows in nearly inaccessible sites on the western tips of two ridges on the island of Kaua'i.

entirely clear whether these animals eat the plants—which would not be surprising— or simply do their damage by degrading the habitat, which is steep and crumbly, caus- ing land slides.

It is a common practice in the biological sciences to name newly described organisms in honor of some individual, presumably in recognition of their contribu- tion to the enterprise in one way or another. We saw earlier, for example, that *Banksia* was named after the naturalist on Cook's first voyage, Joseph Banks. Other examples include *Darwinia*, a genus in the eucalyptus family requiring no further comment, and *Hillebrandia*, a native Hawaiian genus related to begonia, named after Wilhelm Hillebrand who lived in the islands for many years and produced the *Flora of the Hawaiian Islands* in 1888. The genus *Dubautia* recognizes the contributions of J. E. Dubaut, an officer in the French Royal Marines. Lest one gets the idea that recognition in this way honors only those of excellent character and significant accomplishment, a few words about Charles Wilkes, the commander of the U.S. Exploring Expedition of 1838-1842, are in order. His name, of course, has been memorialized as *Wilkesia*. I will quote an entire paragraph from David Stoddart's con- tribution to MacLeod and Rehbock's (1994) *Darwin's Laboratory*. His description is simply too good to paraphrase: "The U.S. Exploring Expedition of 1838-1842 was the first coordinated investigation of the Pacific organized by the government of the United States. Its achievements in hydrographic surveying and in science are right- ly celebrated. But anyone who has ever been on a scientific expedition at sea will rec- ognize it as the ultimate nightmare. It began with years of controversy, was executed through four years of malice, ineptitude, and frustration, and ended with decades of animosity and recrimination. Many if not all of its difficulties must be attributed to its commander, Charles Wilkes, a man of shallow intellect, meager understanding, and emotional instability. Even before the expedition began he was seen as 'exceedingly vain and conceited.' At times during the expedition his very sanity seemed in doubt; he was, as Stanton comments, a leader of a seagoing expedition who was neither leader nor seaman. **At his subsequent court-martial** [emphasis added] he was variously described as violent, overbearing, and insulting; incoherent and rude; and offensive in the highest degree. In the annals of exploration there can have been few leaders more crass, foolish, and inept than Wilkes. The commander of the *Beagle*, by comparison, who was far from the easiest of men, appears positively benign by comparison."

Contemporary botanists have at their disposal a wide variety of research tools that enable them to probe relationships among groups of organisms. The array of techniques run the gamut from classical studies of comparative morphology to the newest methods of gene sequence analysis. The interested reader will find a short description of some of these methods in Appendix One. Several of the examples below include discussions of ancestral relationships of island species based on results using an assortment of those techniques. The first example takes us back to the silversword alliance.

The first suggestion that a relationship existed between *Argyroxiphium* and a group of mainland plants called tarweeds, was that of Asa Gray of Harvard University who, as the reader may recall, was involved in naming some members of the genus in the first place. Gray's view was accepted by other botanists of the time, likely in view of his preeminence in the field as well as the fact that few other workers had ever had the opportunity to study these plants in detail. In 1936 David Keck, a noted expert on mainland tarweeds at the Carnegie Institute on the Stanford University campus, published a detailed reappraisal of the Hawaiian silverswords in which he argued that Gray was wrong in associating the silverswords with tarweeds. Three factors played roles in his position, isolation of the islands, degree of endemism (percentage of native species), and the perilous nature of many of the species. The distance separating the islands from the mainland offered an almost unimaginable barrier for a plant colonist to cross. Therefore, the reasoning went, the source of colonists must have been much closer at one time; perhaps a larger land mass had existed nearby; or perhaps the present high islands are the mountainous remnants of a larger land mass, perhaps even a continent now subsided. Secondly, the high level of endemism in the islands suggested a flora sufficiently old to have had time to accumulate that many unique species. Thirdly, the fact that so many species seemed to be on the edge of extinction was also taken to indicate an ancient flora.

This view prevailed more or less unchallenged until the late 1950s when Sherwin Carlquist (1959), then at the Rancho Santa Ana Botanical Gardens in Claremont, California, presented anatomical and other evidence that clearly pointed to a close relationships between the silversword alliance and mainland tarweeds, and that together they comprise the subtribe Madiinae (named after *Madia*, one of the mainland genera). More recently, DNA sequence studies by Bruce Baldwin and colleagues (Baldwin and Wessa, 2000 and references therein) at the University of California, Berkeley, supported the uniformity of the subtribe and clearly indicated that a colonist from the mainland had indeed been responsible for the introduction of the group into the islands. Who the colonist might have been was the next logical question.

As you will recall, one of the things that one looks for in trying to establish relationships between sets of organisms is their capacity to cross with each other, which reflects upon the degree of divergence that exists between them. The use of crossing behavior had already played an important part in judging relationships within the silversword alliance where all attempted intergeneric crosses had resulted in viable offspring (Carr and Kyhos, 1981, 1986). Crossing experiments between members of the silversword alliance and mainland tarweeds were attempted but were met

with failure until it was realized that another factor had to be taken into consideration. Members of the silversword alliance are tetraploids so it seemed sensible to look for a likely mainland candidate from the tetraploid mainland tarweeds. It was not until diploid tarweeds were tested did successful crosses result (Carr et al., 1996). A measure of a successful cross is the viability of the hybrid individuals. One way to judge this is by the viability of the hybrid's pollen grains. In the case

Plate 38. *Carlquistia muirii*, a Californian member of the mainland tarweeds and an ancestor of the Hawaiian silversword alliance. Photo by Bruce Baldwin.

of some of the island X mainland crosses, viable pollen counts as high as 49% were observed. The mainland plant thought to represent the most closely related ancestral type is *Carlquistia muirii* (Plate 38). The name of this species recognizes the great accomplishments of two Californians, Sherwin Carlquist for his monumental contributions to island biology and John Muir, who some call the father of the American conservation movement.

The historical scenario accounting for the formation of the island species was thought to have begun by two diploid mainland species crossing to form a new, tetraploid species. This is not an unreasonable suggestion considering that several present day continental tarweeds are tetraploids. A seed from the newly formed mainland tetraploid made the chance journey to the islands, probably attached to a migrating bird. In due course, the mainland tetraploid species died out. Arrival of the propagule on one of the Hawaiian Islands, possibly Kaua'i, was followed by germination and successful establishment. The rest, as they say, is botanical history.

The next set of examples continues on the theme of examining evolutionary relationships between island and mainland Californian species using experimental approaches. Some of studies captures information from proteins, while DNA sequence analysis played an important role in the rest of the examples. Taxa are listed alphabetically except for the mints which are listed under that name.

Bidens—*ko'oko'olau, nehe,* beggars' ticks, Spanish needle

Bidens is a genus of perhaps 250 species in the sunflower family (Asteraceae) that is well represented in the Americas, Africa, and Polynesia with a few species in Europe and Asia (*Manual*). Several weedy species are widespread in the world including the Hawaiian Islands. The genus gets its name from two bits of Latin, "bi" meaning two, of course, and "dens" meaning teeth referring to the two barbed awns borne on the

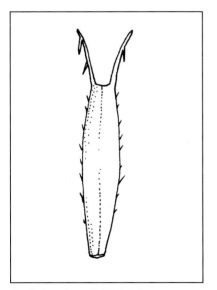

Fig. 2. An achene (fruit) of *Bidens hillebrandiana* subsp. *polycephala* [after K. Nagata and F. Ganders in Lyonia 2: 3 (1983)]. Note the barbed teeth. The body of the achene is a little over 1 cm in length.

achene (fruit). Figure (2) shows an artist's representation of this feature, well known to anyone who has walked through weedy fields in the autumn and picked up these opportunistic travelers. Along with four naturalized species, 19 species of *Bidens* unique to the islands are recognized in the Hawaiian flora. There is strong evidence to suggest that the natives are all derived from a single colonization event. The species are fully interfertile, but hybridization is uncommon in Nature owing to the species' different ecological associations, a situation much akin to the silverswords. Plate (39) features *B. sandvicensis* subsp. *confusa* growing along the Iliau Loop Trail on the rim of Waimea Canyon, Kaua'i. Other specimens of *Bidens* can be seen within the loop area and along the Kukui Trail into the canyon. Farther along the road leading to Koke'e one can often find one of the naturalized species of *Bidens*, *B. pilosa*, easily distinguished by its white ray flowers, opposed to the yellow ray flowers that characterize the native species in this area. *Bidens alba*, another naturalized species, also has white ray flowers, but it has not been reported on Kaua'i according to the *Manual*.

Plate 39. *Bidens sandvicensis* subsp. *confusa* growing on the rim of Waimea Canyon at the Iliau Loop Trail area.

One of the most unusual species of *Bidens* on the islands is *B. cosmoides*, illustrated in Plate (40). This species is easily distinguished by its large flower heads, normally about 6 cm in diameter but specimens are known that were 8 cm in diameter. The unusually long anthers, which are exerted beyond the end of the floral tube, are thought to be an adaptation for pollination by birds. The unusual floral structure led to placing this species in its own monotypic section (sect. *Degeneria*) by E. E. Sherff, an authority on the genus. Also, attempts to cross *B. cosmoides* with other Hawaiian species failed

Plate 40. *Bidens cosmoides* showing its characteristic display of floral parts which were thought to have been selected for bird pollination. Photo by Fred Ganders.

thus providing additional evidence for its separate status. More recent work by F. R. Ganders of the Department of Botany of the University of British Columbia—the author of the treatment of *Bidens* in the *Manual*—revealed that the pollen used by earlier workers in their crossing experiments was inviable. Crosses using fresh pollen resulted in viable offspring. As has been shown in other genera of Hawaiian plants, e.g., the silversword alliance, hybrids may be created in the laboratory but are comparatively rare in Nature because different species occur in different habitats.

Prof. Ganders and his colleagues (2000) also studied the evolutionary relationships among the Hawaiian *Bidens* species and possible ancestors using DNA sequence data. Included in their study were species from the Marquesas and an interesting one found on Starbuck Island. Starbuck Island, lying at 5°37'S, 155°55'W, is one of the Line Islands along with Palmyra, Tabuaeran (Fanning), and Kiritimati (Christmas), among others. Several noteworthy observations came from the DNA-based studies, one of which was the homogeneity of the Hawaiian species including *B. cosmoides*, which, as mentioned above, was at one time thought to belong to a different subgroup within the genus. This observation suggests that *Bidens* has been present on the Hawaiian Islands for a comparatively short period of time during which no differentiation has occurred, at least with regard to the particular gene studied. Hawaiian and Marquesan species were closely related to each other and in turn this group was most closely related to a group of species from Central America. The "Starbuck Island *Bidens*," however, was not part of the assemblage of Polynesian species suggesting that it is the result of an independent colonization event. Its closest relatives have not yet been established. This species has been formally named *Bidens kiribatiensis*.

Gossypium—cotton and relatives on the islands

Malvaceae, the family to which cotton belongs, include several other species that may be familiar to most readers, hibiscus (*Hibiscus* species), hollyhock (*Alcea rosea*), the original marshmallow (*Althaea officinalis*), and okra (*Abelmoschus esculentus*). The tree mallow (*Lavatera* species) is also commonly planted as a decorative garden plant. Malvaceae are a moderately large family comprising 1,800 species in over 100 genera. The Hawaiian flora includes 49 species representing 16 genera. Two of the genera, *Hibiscadelphus* and *Kokia*, are native to the Hawaiian Islands; we will learn a little more about them later.

Cotton belongs to the genus *Gossypium*, which is probably one of the most intensively studied groups of plants due in no small part to its being one of the oldest cultivated plants known. The characteristic that makes a few of these species so valuable is the presence of very long hairs on the seeds, which, when dried, are capa-

Plate 41. *Gossypium tomentosum,* the endemic Hawaiian cotton. Photo by Ken Marr.

Plate 42. *Gossypium darwinii,* the endemic cotton of the Galapagos Islands.

ble of being spun into thread. The history of development of cotton as a crop and the relationships between Old World and New World species have been subjects of extensive biochemical, cytological, ethnobotanical, genetic, and morphological studies. The literature generated by this work is vast and can only be touched upon in the briefest way here. A little background, however, is necessary to put the reader in the picture.

The genus consists of about 50 species native to warm temperate and tropical regions. Six species of cotton occur in the New World; all are tetraploids ($2n$ = 52) having originated by hybridization between diploid parental species ($2n$ = 26) at some time in the distant past. The two most important cultivated species of cotton, G. *barbadense* and G. *hirsutum*, are both tetraploids. These species occur over moderately wide ranges with G. *barbadense* centered in northern South America and the Caribbean and G. *hirsutum* common in Central America and the Caribbean region but extending into northern South America. The remaining four tetraploid species occur over much smaller ranges: G. *mustelinum* occurs in northwestern Brazil, G. *lanceolatum* in southwestern Mexico, G. *tomentosum* (once called G. *sandvicense*) in the Hawaiian Islands (Plate 41), and G. *darwinii* in the Galapagos Islands (Plate 42).

Three species of cotton occur in the Hawaiian Islands, the native species that we have just met, known to Hawaiians as ma'o and *pulupulu*, which I have seen translated as "hairy-hairy." Among other unrelated meanings, *pulu* refers to the shiny golden-brown wool-like material that occurs on the stems of tree ferns belonging to the genus *Cibotium*. The native cotton occurs on dry sites on all of the main islands except Hawai'i. Two species were imported to the islands for the purpose of cultivation, G. *barbadense*, the sea island cotton or *pulupulu haole* (literally, foreign cotton), and G. *hirsutum*, commonly known as upland cotton. Cultivation, however, is no longer carried out on the islands.

Answering the question as to the origin of the Hawaiian species is compli-
cated by the lack of a clear-cut similarity with any of the other tetraploids, which is
a situation frequently encountered in studies of Hawaiian species that have diverged
in many ways from their continental ancestors. Adding to the problem was the fact
that different sets of characters, e.g., some floral and fruit features, nature of some
modified leaves (technically, bracteoles), leaf indumentum (furriness), and certain
pigments, each pointed to a different possible relationship within the tetraploid
species. A study of chromosomal features also failed to establish the most likely
ancestral species. The ease by which G. *tomentosum* and continental species crossed
to yield hybrids also indicated that there had been comparatively little genetic differentiation between them. The application of DNA sequence data from several
genes, however, provided reliable support for G. *hirsutum* as the most likely ancestral
species. Since G. *hirsutum* occurs over a considerable area it was of interest to see if
any genetic clues could be found that pointed to a specific area within that range
from which the Hawaiian species originated. A search for such clues failed to pin-
point a specific region which led the workers to suggest that the origin of the island
species occurred during an early stage of diversification of G. *hirsutum*. It is also interesting to see that, while results from chloroplast DNA analysis did not resolve the
species relationship problem on its own, they did provide an important clue as to the
possible time of colonization. DNA sequence divergence values determined for several species, diploid and tetraploid, revealed that the tetraploid line of cotton originated between one and two million years ago, well within the time during which
Hawaiian Islands existed (DeJoode and Wendel, 1992).

It seems very likely that transoceanic dispersal of a seed must have occurred.
The findings that cotton capsules are buoyant and that seeds remain viable for periods of up to three years in artificial seawater clearly indicate the likelihood of colonization of the Hawaiian Islands by flotation. It is interesting to note that several
other species of cotton have distributions that suggest oceanic transport and, fur-
thermore, that they tend to occur in coastal or strand vegetation.

Hibiscus, well known in the horticultural trade, also has native representatives on the Hawaiian Islands. Illustrated in Plate (43) is *H. waimeae*, known to the
Hawaiians as *koki'o ke'oke'o*, (lit., white hibiscus) was photographed in the Limahuli
Gardens on Kaua'i. I have also seen it used as a walkway border planting in hotel
grounds, also on Kaua'i. This species occurs naturally on Kaua'i and, as its name suggests, can be found in Waimea Canyon, but it also occurs in valleys along the northern and western coasts of the island. A closely related species, *H. arnottianus*, occurs
on O'ahu and Moloka'i. Of the 10 species of *Hibiscus* listed in the *Manual*, five are
native and listed as either rare or endangered. *Hibiscus tiliaceus*, known to the

Hawaiians as *hau*, occurs widely in subtropical and tropical regions. There is some question as to whether the Polynesian settlers may have brought *hau* with them; this would not be surprising knowing its range of occurrence and overall usefulness. On the Hawaiian Islands it occurs on all of the main islands and on Midway and the French Frigate Shoals. *Hau* is a medium sized tree many parts of which were utilized by the Hawaiians for such items as spars for outrigger canoes, sandals, cordage for house building and other jobs where a stout cord was required, and for some medicinal preparations (Krauss, 1993; Kepler, 1998).

Plate 43. *Hibiscus waimeae*, a Hawaiian endemic hibiscus, native to Kaua'i, but frequently grown as a decorative plant.

Other members of the family native to the Hawaiian Islands include the genera *Hibiscadelphus* and *Kokia*. These genera represent other examples of the perilous, if not outright extreme, condition of native taxa that we have seen before in this book. *Hibiscadelphus* consists of six species, three of which are extinct with the remaining ones listed as endangered. In this instance, as in many others, endangered is likely an understatement. Of the three remaining species the most abundant in the wild would appear to be *H. distans*. Thought at one time to have disappeared, 10 individuals were found in 1972, with the number jumping to 16 as a result of a return visit to the site near Waimea Canyon on Kaua'i. The authors of the *Manual* express some doubt about the well-being of the species following the 1982 hurricane ('Iwa), however. *Hibiscadelphus giffardianus* was identified from a single tree in 1911, and is likely finished in Nature although it is maintained in cultivation. Two or three individuals of *H. hualalaiensis* were known in the wild in 1977, but this species is known now only in cultivation.

This perilous situation exemplified by *Hibiscadelphus* offers us an opportunity to look at the relationship that exists between some plants and their pollinators. Open-faced flowers, daisies for example, present little in the way of challenge for potential pollinators. They fly in, collect some nectar or pollen, and fly off to the next flower. This could be called landing-field pollination for lack of a technical term. If plants have tubular flowers, however, the scene changes dramatically. Big insects can't get into small tubes, and small insects may be lost in large ones. Evolution has resulted in flowers that are adjusted to fit a given group of insects, while insects have adaptations that suit them for particular types of flowers. These adaptations between floral structure and visitor lead to maximum efficiency of visits and pollen transfer. This is co-evolution at work. *Hibiscadelphus* has (had?) a similar problem. The flowers of these species are curved requiring a pollinator capable of reaching the bottom of the floral tube for its nectar reward. With the disappearance of its curved-bill pollinators, the honey-creepers, however, this group of plants was at serious risk. The next obvious question is why did the honey-creepers disappear? According to Jared Diamond, a bird biologist of international fame, *Hibiscadelphus* was the victim of "trophic cascade." This is a technical term that most of us would know simply as the domino effect: do something harmful at one place in the food chain and something farther along the path will be affected. Familiar examples of this phenomenon are the effect that DDT, used to kill mosquito larvae, had on birds (their eggs, actually), and the movement of polychlorinatedbiphenyls (PCBs) through the food chain. According to Dr. Diamond, *Hibiscadelphus* species died of malaria, not directly, of course, but because their pollinators, the honeycreepers, had been bitten by mosquitoes carrying avian malarial parasites. Malaria killed the birds, dead birds don't pollinate, and unpollinated flowers don't set seed. A detailed discussion of the biological domino effect, and more information about Jared Diamond's ideas, can be found in David Quammen's (1996) marvelous book, *The Song of the Dodo*.

The situation with respect to extinction in *Kokia* is much the same; one of its four species is extinct and the other three are considered endangered. Two of the endangered species are being maintained in cultivation, one, *K. cookei*, by grafting onto *K. kauaiensis*. Efforts some years ago to replant *K. cookei* on Moloka'i and O'ahu failed. It seems likely that none of the species in either of these genera was ever particularly abundant, which seems to be a common situation in the Hawaiian flora, as mentioned in other places in this book.

Two other genera in Malvaceae represented in the islands by native and/or indigenous species, as well as naturalized ones, are *Abutilon* and *Sida*. The three native species of *Abutilon* are classified as endangered. Fairly common on all of the islands and on Midway is *A. grandifolium*, known commonly as hairy abutilon and by the Hawaiians as *ma'o*, a word also used to identify the native cotton species. This is a widespread tropical weed that appears in dry areas, often in abandoned fields. The specimen that appears in Plate (44) was photographed beside the road to Polihale

Plate 44. *Abutilon grandifolium*, a naturalized member of Malvaceae (cotton family) commonly found in waste places such as deserted cane fields.

Plate 45. *Sida fallax*, an indigenous species in the islands frequently used for making *lei*.

State Park (western coast of Kaua'i) on margins of an abandoned sugarcane field. This example could, of course, been included in Chapter Three along with other weedy introductions.

Sida is represented on the islands by seven species, one, or possibly two, of which are indigenous; the others are all uninvited guests. Plate (45) features *S. fallax*, 'ilima in Hawaiian, one of the indigenous species. The photograph was taken at the Iliau Loop Trail (Kukui Trail trailhead) on Kaua'i where the brightly colored flowers stand out clearly. This species can survive on a variety of substrates as witnessed by its occurrence on limestone reefs on Midway Atoll, and in a variety of forest types on most of the islands. The flowers are popular for making *lei*. In a totally non-botanical context, Douglas Pratt (1998) notes in his *Pocket Guide to Hawai'i's Trees and Shrubs* that two taxi cab companies in Honolulu have adopted the Latin (*Sida*) and Hawaiian (*Ilima*) names for their businesses. So far as I can recall, I have only used the service of the former, although I shall look for the latter cabs on my next visit to avoid favoritism.

Lepidium—pepperwort or peppergrass

Similar conclusions were reached concerning the evolutionary history of *Lepidium* by a group of workers who also employed DNA sequence differences as a source of data. *Lepidium* is a member of the cabbage family (Brassicaceae, Cruciferae) represented in the Hawaiian Islands by two native species (*L. arbuscula* and *L. serra*), another that is widespread in the Pacific Basin (*L. bidentatum*), and five weedy species that have become naturalized. Although the results of the study (Mummenhoff et al., 2001) provided no insights into the relationships among the island native species, an origin from a Californian ancestor was clearly indicated. The differences in DNA sequences between island and mainland groups of species suggested that *Lepidium* has been present on the islands for about 350,000 years.

Hawaiian lobelia relatives

If I were asked to judge which group of Hawaiian plants presented the most spectacular botanical show, limiting the choice to native and indigenous species, I would have difficulty choosing between the silverswords (*Argyroxiphium* and *Wilkesia*) and the subject of this next discussion, the Hawaiian lobelia relatives. I must confess that I would take the easy way out and award them both blue ribbons, the swords for their stunning floral display, and the lobelioids for their rich variation on a theme. In this

section, I will try to give some idea of the richness of this group, as well as use the opportunity to talk a little about their fascinating evolutionary history. But first, some background and a few numbers.

Lobelia (Plate 46), a common garden flower frequently used as a border plant, belongs to Campanulaceae a moderately large family consisting of 70 to 80 genera and about 2,000 species. The family derives its name from *Campanula*, the bell flower. The family is further divided into two subfamilies, one of which is based upon *Campanula*-like flowers and the other upon *Lobelia*-like flowers. The species that are of interest in this section belong to the second group, hence the use of the term

Plate 46. A cultivated lobelia.

"lobelioid." As pointed out by Tom Lammers in his discussion of the group in the *Manual*, the Hawaiian lobelioids were described by Hillebrand (1888) as the "peculiar pride of our flora." What makes this group so special is the existence of no less than six native genera, which, along with 13 native species of *Lobelia* itself, contribute 110 native species to the Hawaiian flora, the largest group in the islands in terms of number of native species. The native genera, with total number of species and the number of extinct species in parentheses, are *Brighamia* (2), *Clermontia* (22, 1), *Cyanea* (52, 14), *Delissea* (9, 7), *Rollandia* (8, 2), and *Trematolobelia* (4). Of the 13 species of *Lobelia* native on the islands, two are thought to be extinct.

Brighamia (*ālula, hāhā*) is certainly one of the most peculiar plants that one is likely to see in the islands—sort of an oceanic boojum (*Fouquieria columnaris*, a pyramidal plant from the deserts of Baja California, belonging to Fouquieriaceae). To see *Brighamia* in its natural setting, however, would be no small fete. Both species occur on virtually inaccessible cliffs. *Brighamia insignis* occurs on cliffs on the Nā Pali Coast of Kauaʻi, and formerly on Niʻihau, where it was last seen in 1947. *Brighamia rockii* occurs on the sea cliffs of Molokaʻi and is thought to have occurred on Maui and possibly Lānaʻi in the past. A study of genetic variation in the two species by Gemmill et al. (1998), based on about 41% of all known individuals in the wild—an impressive level of sampling!—revealed that the two have significant differences in their electrophoretic behavior within the set of proteins studied. This is somewhat unusual considering the very close similarities exhibited by other morphologically similar sets of species in the Hawaiian flora. This study provided background for attempts to generate larger numbers of individuals for replanting in Nature.

Brighamia insignis can be seen without difficulty at the Kīlauea Lighthouse on the northeastern coast of Kauaʻi, where it is being cultivated for reintroduction, at the Limahuli Botanical Garden on Kauaʻi's north coast, and along the Wailea Point Trail, Wailea, Maui (southwestern East Maui). *Brighamia* is also available from native plant nurseries as well. The photographs for Plates (47) and (48) were taken at the Kīlauea Lighthouse site. *Brighamia*, which incidentally was named after W. T. Brigham, the first director of the Bishop Museum in Honolulu, is the subject of an article by D. P. Hannon and S. Perlman in the April-May 2002 issue of the *Cactus and Succulent Journal*. The cover features a photograph of *B. rockii* taken by S. Perlman, a well-known field botanist in the Hawaiian Islands.

Clermontia (*ʻōhā wai, ʻōhā, hāhā*) with its 21 extant species, consists of branched shrubs or small trees; some species can be epiphytes. The branched growth form of this native genus distinguishes it from the other genera in the island lobelia group, many species of which are unbranched, often resembling small palm trees. *Clermontia* exhibits an interesting distribution pattern in the islands with the highest level of endemism seen in the youngest island; nine of the 11 species found on the island of Hawai'i occur nowhere else. Maui has 10 species of which only two are native to that island. Species numbers and endemism drop off as one travels to the oldest island, Kaua'i, where one finds only one species of *Clermontia*, which is also found on O'ahu. The sole species on Kaua'i's is C. *faurei*, which can be readily found growing beside the Pihea Trail from about the one mile marker onward. This species bears the characteristic features of the genus with the branching shrub growth form. Flowers, which typically emerge from the leaf axils, are dark purple as seen in Plate

Plate 47. *Brighamia insignis* growing under cultivation for eventual reintroduction.

(49). Also characteristic of this genus is the formation of orange fruits that have a faint resemblance to small pumpkins. Plate (50) features an example of C. *arborescens*, where the much branched growth form can be seen. This specimen, which was in bud, was photographed in the boggy area along the Boy Scout Trail on West Maui. It also occurs in similar habitats on Moloka'i and Lāna'i. The very beautiful C.

Plate 48. The *Brighamia* "farm" at Kīlauea Lighthouse site on the north coast of Kaua'i.

Left:
Plate 49. *Clermontia faurei*, an endemic species known only from rain forests of Kaua'i and O'ahu. It is the only member of the genus on Kaua'i.

Above:
Plate 51. *Clermontia micrantha*, a species limited to bogs of Lāna'i and West Maui. Photo by Ken Marr.

Plate 50. *Clermontia arborescens* growing in a boggy area along the Boyscout Trail on Maui. Note the bushy growth form.

Plate 52. *Clermontia parviflora* photographed near the entrance to the Thurston Lava Tube. Photo by M. Hawkes.

micrantha (Plate 51) occurs in bogs and wet forests and is common on Pu'u Kukui, the high point on West Maui, and in nearby 'Eke Crater. A species that is easier to find is the attractive C. *parviflora* (Plate 52), which was photographed along the paved walkway leading to the Thurston Lava Tube. This is one of the species that occurs only on the island of Hawai'i where one finds it in wet forests.

One of the largest native genera in the Hawaiian Islands is *Cyanea* (*hāhā*) with 52 species, some of which are unbranched, while others are described in the *Manual* as branched or sparingly branched. Unfortunately, much of the variation within this group of fascinating plants can not be appreciated because more than a quarter of them no longer exist. Also, several of the surviving species are listed as endan-

Plate 53. *Cyanea pilosa* subsp. *longipedunculata* showing typical orange fruits. Photo by Ken Marr.

Fig. 3. *Cyanea hardyi* growing in the Limahuli Botanical Garden on Kaua'i. Note the wavy leaf margins and pendulous flowers.

gered. A spectacular member of this genus, C. *hardyi*, with its long wavy leaves and numerous pendulous flowers, is rare in Nature but is being successfully maintained at the Limahuli Botanical Garden. Figure (3) is an artist's illustration of this species.

Cyanea pilosa is a medium size shrub that occurs in wet forests on Mt. Kohala and the windward sides of Mauna Kea and Mauna Loa. Pictured in Plate (53) is C. *pilosa* subsp. *longipedunculata* in fruit. This subspecies occurs on the two southern mountains (subsp. *pilosa* occurs on Kohala and Mauna Kea).

Delissea is listed in the Manual as consisting of nine native species, but most of them are thought to be extinct, some of these known only from the original collections. The only plants of this species that I have seen are under cultivation in the Limahuli Gardens. Otherwise, D. *rhytidosperma* is known from mesic forests on Kaua'i. The photograph for Plate (54) was taken in the Gardens.

Plate 54. *Delissea rhytidosperma*, one of the remaining members of this Kaua'i native genus, photographed in the Limahuli Garden.

Plate 55. *Lobelia dunbariae* ssp. *paniculata*, a true lobelia that occurs only on Moloka'i. Photo by Ken Marr.

Lobelia is represented on the Hawaiian Islands by 14 native species, two of which are thought to be extinct. *Lobelia dunbariae* subsp. *paniculata*, illustrated in Plate (55), occurs on Moloka'i where it grows on wet forested cliffs. Species of *Lobelia* from the Hawaiian Island can be sorted into two groups based upon morphological characteristics which has led to the suggestion that two independent colonizations may have occurred, possibly one from Asia and a second from South America.

After an all too brief survey of this magnificent assemblage of plants, it's time to look at their origins and relationships. It is not surprising that early botanists suggested that this assortment of genera must have originated by several independent colonizations, possibly one to account for each genus. The obvious similarities among some of the genera, however, gave rise to some speculation that differentiation could have occurred on the islands after the original colonists (still plural) had become established. The idea that all of the native genera developed on the islands after a single event did not seem a very likely possibility. Recent research using DNA sequence analysis has shed new light on this system, however, suggesting a totally different scenario. Thomas Givnish at the University of Wisconsin and colleagues (1995, 1996a,b) examined DNA of 76 species representing all native genera of Hawaiian lobelioids. A diagram of the relationships suggested by these data is shown

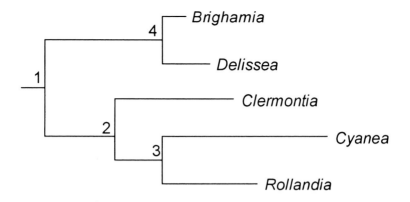

Fig. 4. Hawaiian lobelioid relationships indicated by studies of DNA sequence data. The fundamental conclusion drawn from these studies is that all Hawaiian members of this group, excepting *Lobelia* species, arose from a single introduction some 15 million years ago. The numbers indicate the order of branching in the evolution of the genera. Each number refers to a common ancestor shared by genera that lie farther to the right in a particular genealogy, e.g., No. 1 represents the ancestor that gave rise to all current genera; No. 3 represents an ancestor common to *Cyanea* and *Rollandia*.

in Figure (4). This kind of diagram, called a "phylogram," is one of the ways used to identify "sister groups," genera in this case. Sister groups are groups of organisms that are likely to have shared a common ancestor. The methods used yield diagrams where the lengths of the respective arms are proportional to the number of mutations detected in the DNA. A higher number of mutations, of course, would be taken to indicate a longer period of time and thus, a larger degree of divergence between two lineages.

The most remarkable outcome of this analysis was the apparent origin of all of the native genera from a common ancestor, and therefore, a single colonization event! Also surprising was the finding that *Brighamia* fits into the scheme, which seems contrary to the suggestion that it is unique within the natives and therefore arrived independently. The suggestion that *Brighamia* is not part of the other apparently closely related genera was based upon fruit structure. Fruits of *Brighamia* are hard, dry capsules, in contrast to the fleshy fruits of the other native genera. A closer examination, however, reveals that the fruits of *Brighamia* pass through a fleshy stage in their development. Dry capsules are a secondary difference. This example serves well to emphasize the need to look at as many aspects of a plant's life and as many structural features as possible in the search for clues as to its relationships.

The reader may recall from Chapter One that the lobelioids in the Hawaiian Islands are among the oldest plant lineages for which reliable data are available. At about 15 million years of age, the ancestor of this amazing group of organisms would have had to establish a foot hold on a high island that no longer exists. At that time the most prominent island would have been Gardner. Five million years later (10 million years before the present) Gardner would have still existed, but it would have been transported further westward, and would have been reduced significantly in size. At ten million years before the present, however, Necker Island would have been over the hot spot, and could have provided sites for colonization by lobelioid ancestors. The key point to all this is that there would have been sites available for island hopping that could have provided ancestors of the present lobelioids access over time to the current high islands.

Metrosideros—'ōhi'a lehua

Myrtaceae, the myrtles, are a large family consisting of about 140 genera with the number of species estimated at about 3,000 by some writers (the *Manual*) or even higher by others. The most familiar members of this family are the eucalypts or gum trees (*Eucalyptus*)—the vast majority of the 600 or so species of which are Australian natives. Over 90 species of eucalypts have been planted in the Hawaiian Islands with about 30 that are naturalized. Several other genera in this family have been introduced to the islands, including *Rhodomyrtus*, represented by *R. tomentosa*, the rose myrtle (Plate 56), and two species of guava (*Psidium*), *P. guajava*, the guava of commerce, and *P. cattleianum*, the strawberry guava, have become nuisances and will be described in more detail in Chapter Three.

Of interest to us here, however, is the genus *Metrosideros*, which consists of about 50 species, the bulk of them native to New Zealand and other islands in the southwestern Pacific Ocean. A few other species are found on some of the high islands in the southern Pacific Ocean, and, interestingly, a single species occurs in South Africa. The Hawaiian Islands are home to five native species, one of which, M. *polymorpha*, is among the most widespread trees on the archipelago and, as its specific epithet suggests, it is highly variable. The level of variation within this species has been accommodated by recognizing eight varieties. It is no exaggeration to suggest that some experience

Plate 56. *Rhodomyrtus tomentosa*, a naturalized member of the eucalypt family (Myrtaceae) referred to as the downy or rose myrtle.

Left:
Plate 57. *Metrosideros polymorpha,* perhaps the most variable of all Hawaiian native species.

Above:
Plate 58. Flower color in *'ōhi'a* ranges from the very common red through pink to nearly white.

with this group is required for ready identification! The species in general, however, is readily recognized by its tufts of bright red flowers, as seen in Plate (57), but pink, white, and yellow color variants are known. Plate (58) illustrates the extremes in flower color. The Hawaiians call these plants *'ōhi'a lehua,* or simply *'ōhi'a.* Because the genus has been studied in some depth, it is possible to put the likely origin of the Hawaiian species into a broader context. We can do that next.

The center of origin of *Metrosideros* is thought to have been New Zealand,and it is from those islands that colonists originated from which other concentrations of species have arisen. Based upon DNA sequence data, S. D. Wright and coworkers (2000, 2001) suggested that the present array of species could be accounted for by three radiations, one that gave rise to species in New Caledonia, with subsequent colonization of the Bonin Islands (south of Japan), Fiji, and the Solomon Islands, and two from which the other oceanic species arose. The older of these latter two events, which resulted in getting *Metrosideros* to Samoa, Fiji, and Vanuatu is thought to have occurred possibly as long ago as two million years. The third oceanic dispersal, resulting in colonization of Rarotonga, Tahiti, Lord Howe Island, the Kermadec Islands, Pitcairn Island, the Hawaiian Islands, and the Marquesas occurred much more recently, perhaps as recent as the Pleistocene glaciation (that is, within the past two million years). The big surprise from this study (the 2001 paper) was that *Metrosideros collina* from the Marquesas and M. *polymorpha* from the Hawaiian Islands are more closely related to each other than to any other species. The differences in sequences indicated that introduction of *Metrosideros* to the Hawaiian Islands is likely to have occurred between a half and one million years ago. This date is in reasonable agreement with sub-fossil remains of *Metrosideros* on the Hawaiian Islands that have been dated at about 350,000 years. Differences between *Metrosideros* on the Marquesas and other species farther west suggest a residence time on the Marquesas of two million years before the jump to the Hawaiian Islands occurred.

Scientists usually look for the simplest explanations of phenomena, the most parsimonious path from observation to explanation. In the case of *Metrosideros* on the Hawaiian Islands, a much simpler path than one involving the Marquesas would have brought the genus to the islands via Samoa, a route that had been suggested by earlier workers. The problem with the Marquesas as a source is that they are nearly 3,000 km from the Hawaiian Islands (south and a bit to the east) with no intervening high islands that could serve as stepping stones. In addition to distance, there is another barrier, something called the inter-tropical convergence zone (ITCZ), which lies between 5 - 10°N latitude and marks the discontinuity between the hemispheres. This zone separates the southern and northern hemisphere air circulation patterns and provides a formidable hurdle for movement of airborne seeds. In simpler terms, still, winds in the southern hemisphere blow in the wrong direction for them to be invoked as a way to get seeds carried more or less directly to the Hawaiian Islands. From time to time, however, in the late summer and winter in the southern hemisphere, the ITCZ forms below the equator between 5 - 10° S latitude. This shift touches the Marquesas but does not go as far south as Tahiti. Seeds lofted from the Marquesas would be carried northward by this high altitude air flow system, which descends in the vicinity of 25° N latitude. Propagules could then be picked up by the trade winds and delivered to the islands from the northeast. This might be likened to a municipal transit system where long distance travel requires a transfer. As Wright and his coworkers (2001) pointed out, it would be very interesting to see if other Hawaiian species might have come from Marquesan ancestors by this route.

One other requirement for this path to be successful has to be met, however, and that has to do with whether the species in question have seeds that are up to the rigors of the trip. According to Carlquist (1980) *Metrosideros* is one of a short list of plants whose seeds appear to meet the requirements. The seeds are small and light and are viable at -30°C for at least six hours and can tolerate sea water for at least a month.

Mints

I imagine that almost everyone reading this is familiar with the mint family (Lamiaceae or Labiatae) in one form or another. Spearmint and peppermint (*Mentha* species) are flavoring agents used in many commercial products including candies and toothpaste; lavender (*Lavandula* species) is widely used in perfumery and soaps. Many familiar kitchen herbs are also members of the mint family, these include basil (*Ocimum*), marjoram and oregano (*Oreganum*), rosemary (*Rosmarinus*), sage (*Salvia*), and thyme (*Thymus*). Another mint species with a well known effect is *Nepeta cataria*, commonly known as catmint or catnip (a friend of mine calls it kitty pot). All members of the cat family find this plant irresistible. All of these plants have glands on their leaves that contain the aromatic oils. That not all members of the mint family have the capacity to produce these compounds will be discussed further below.

Before we get into the possible evolutionary significance of these compounds in the Hawaiian native mints, it is useful to learn a little about the species that occur on the islands, and something about those that no longer occur on the islands, or anywhere else. Indeed, the Hawaiian Islands have their fare share of mints, some of which are part of the alien flora and represent escapes from gardens, others are uninvited travelers that seem to make their way to available sites almost anywhere in the world, and some are found nowhere else. A few of the common visitors were mentioned above, e.g., basil, horehound, and spearmint. Some of these are spreading, but what their ultimate impact on the native flora will be remains to be seen.

The Hawaiian Islands are also home to *Lepechinia hastata*, sometimes referred to as a pitcher sage. Members of this genus are known from South America, Mexico, California, islands in the Indian Ocean, and, interestingly, Socotra Island, Yemen. Its place in the Hawaiian flora, whether indigenous or introduced, remains an unanswered question, although its presence on other oceanic islands suggests that it may be capable of being transported over long distances by natural means (wind or by birds). Another probably indigenous species is *Plectranthus parviflorus* (small flowered *Plectranthus*, also called spurflower). This plant occurs on other islands in the Pacific basin and in Australia. How it got to the Hawaiian Islands is also not known.

Three genera of mints occur exclusively on the Hawaiian Islands, with a single exceptional species. According to the *Manual*, these genera, *Haplostachys*, *Phyllostegia*, and *Stenogyne*, comprise a total of 52 species, of which 15 are considered to be extinct with several others classified as rare or endangered. It is instructive to examine these genera in some detail in order to appreciate the sensitivity of native species to the combined forces of Nature and human invaders. Of the five species of *Haplostachys* listed in the *Manual*, four are extinct and the remaining one is listed as endangered. *Haplostachys truncata*, one of the extinct species, is known only from the type collection, which was made on Maui in the middle of the 19th century. A word of explanation about "type" collections is in order. When a plant is first discovered in Nature, the collector makes field notes including details of the location, what other plants are growing in the vicinity, and features that might be lost in drying such as flower color or aroma, or possibly what insects might be in residence. In order to validate the name of the new species, a dried specimen is prepared for deposit in a permanent collection. The specimen is referred to as the "type specimen" and serves as a reference to which all other collections of this species can be compared. Similarly, the place where the new species was found is referred to as the "type location." In practice, several specimens are collected for deposition in several herbaria, especially if the plant appears to be new to science, as certainly most Hawaiian species were to the early botanists. There is a risk in this procedure, however, and that is simply that at the time of the discovery, the collector does not know that he (this is not sexist, most early collectors were men) is in the midst of the only population of the plant that exists! He may come away with an excellent set of specimens but has either decimated the species in so doing, or has reduced the size of the population to the point where it can no longer reproduce. Thus, collection of the specimens of *H. truncata* in the period 1851-1855 may well have resulted in the extinction of the species, a serious consequence resulting from a normal field procedure. The authors of the *Manual* suggest, however, that this entire genus may have been rare even at the time of Captain Cook's visit to the islands in 1778. The other three extinct species managed to survive into the 20th century but do not appear to have been able to hold their own in the face of accelerating human activity. *Haplostachys bryanii* was last collected on Moloka'i having been the victim of habitat destruction associated with pineapple cultivation. Similarly, *H. linearifolia* was last collected on Moloka'i in 1928, likely suffering the same fate as its cousin. West Lāna'i was the site of the last collection of *H. munroi*; the year was 1935. Considering the devastation that has been visited on Lāna'i over the centuries (recall the discussion of this phenomenon in Chapter One), it is surprising the species lasted that long. That brings us to *H. haplostachya*, the only extant member of the genus. At one time this species was known from two other islands, Maui and Kaua'i, but at the present it is known only from a single population on the Island of Hawai'i. This species is endangered.

We might pause at this point for a moment and contemplate the extinction of a group of plants, perhaps the genus *Haplostachys* in this instance. It is almost as if we were saying that things were going along pretty well until European botanists showed up and collected specimens willy-nilly without any consideration of the damage they might be wreaking. There is certainly an element of truth in that, but, as the authors of the *Manual* point out, some of these groups of plants may well have been on the brink of extinction at the time the Europeans arrived. In other words, these species may have been at risk solely because of natural factors. Perhaps there never were large numbers of individuals (many Hawaiian native species consist of comparatively few individuals) and that normal fluctuations of population size, perhaps because of low seed yield one season, had taken them to a number of individuals less than the minimum number needed for successful reproduction. Also likely is the possibility that two bad years in a row could have put the species "over the edge." Other factors involved in survival include competition for limited resources and the fickleness of the weather. Clearly, an enthusiastic collector could easily add to the problems faced by a population of plants already under stress. It is also important to realize that the problem was not necessarily one of short term risk. Rather, these organisms could have been on the islands for a considerable period of time, perhaps millions of years, and that they may simply have run out of genetic variability (all species ultimately die). Without maintaining a certain level of genetic variation, species are at risk of accumulating deleterious mutations that can seriously affect survival.

Phyllostegia is somewhat better off. This is a moderately large genus with 27 species known from the Hawaiian Islands with one native to Tahiti (the one exception noted above). Of the 27, seven are extinct, five are considered rare, four are endangered, and 11 are moderately common. It is useful to define the four risk categories used by the authors of the *Manual*: (1) A species is considered extinct if, based on extensive field experience, it is thought to no longer exist in Nature. A few species of Hawaiian plants are extinct in the wild but are maintained in cultivation. (2) An endangered species is one that is at risk of becoming extinct throughout its range unless the threats to it are alleviated. (3) A species is considered vulnerable if it is in immediate risk of extirpation unless the threats are alleviated, or at least reduced; and (4) A species is listed as rare if it is represented in Nature by a comparatively small number of individuals or populations, typical of many island natives, but is not considered at risk **at the present time** (emphasis added). These are necessarily subjective judgments that may move in either direction depending upon new populations being found, failure to find previously known populations, and the appearance of new threats to habitats.

Let's look at the extinct species, as above, again paying attention to when they were last seen. *Phyllostegia knudsenii* is known only from its type locality, which was in the Koke'e area of Kaua'i. Two species collected before 1871, and presumably not seen since, are *P. brevidens*, known from two locations on Hawai'i and *P. hillebrandii*, known from only one location on East Maui. The other four extinct species, along with where and when last collected, are *P. rockii*, three collections on East

Maui, 1912; *P. wawrana*, four sites on Kaua'i, 1926; *P. imminuta*, East Maui and Lāna'i, 1928; and *P. variabilis*, Kure and Midway atolls and Laysan, 1961.

The third genus here is *Stenogyne*, which the *Manual* credits as consisting of four extinct species, five endangered ones, and 11 that appear to be free of risk. The four extinct species are *S. viridis*, which was collected once on West Maui (no date given); *S. cinerea*, collected once on east Maui in 1865; *S. oxygona*, last collected on Hawai'i in 1952 and likely extinct; and *S. haliakalae*, last collected in 1936. A photograph was taken of a single plant of *S. haliakalae* in the Kula Forest Reserve in 1973; the site was revisited in 1984 but the plant was not found.

Several species of *Stenogyne* are comparatively abundant and fairly easy to find. The three with which I have had some experience offer a chance to appreciate both the structural variation present in the genus as well as the different habitats in which they grow. All three are accessible by normally traveled routes. *Stenogyne kamehamehae* (Plate 59) is a vine that grows in wet forests above about 3,000' elevation on Maui and Moloka'i. On Maui the plant produces flowers that are cream-colored, often with a pinkish tinge on the upper lip. Flowers of this species on Moloka'i, which I have not seen, tend to be redder according to the *Manual*. Access to this species on West Maui can be had by way of the Waihe'e Ridge Trail, also referred to as the Boy Scout Trail, whose trailhead is near the entrance to Camp Maluhia a few miles northwest of Wailuku along Route 34. Detailed directions for getting to this trail can be found in Robert Smith's guide *Hiking Maui*. The trail is a bit over two

and a half miles long and terminates on a small rise called Lanilini Peak (literally, "small heaven"). The plants of interest can be found snaking around in the bush about a hundred yards or so before the end of the trail. The plant tends to be a bit scruffy looking but the flowers are large enough to be seen easily. This and other native species can also be seen in the Nature Conservancy's Waikamoi Preserve on Maui.

Plate 59. *Stenogyne kamehamehae*, one of the native Hawaiian mints whose name reveres the line of Hawaiian kings. Kamehameha I united the islands into a single kingdom.

Incidentally, several other interesting plants can be seen along this trail including an abundance of 'ohi'a, many ferns, *Freycinetia arborea* with its very photogenic "cones" (see below), and several members of the Hawaiian

Plate 60. *Stenogyne microphylla* can be found growing entwined in the branches of *Sophora chrysophylla* (*māmane*) at higher elevations on Mauna Kea, among many other places.

lobelia assemblage. Less enthralling is the clear evidence of alien plants along the trail, including dense patches of *Psidium cattleianum*, the strawberry guava (Chapter Three) and, in several places, the invasive *Melinis minutiflora*, the well known and infamous "molasses grass" (Chapter Three).

      Stenogyne microphylla (Plate 60) occurs in subalpine forest habitats on East Maui and on Hawai'i. My favorite spot to see this species is just below the Hale Pōhaku Visitor's Center on Hawai'i at about 9,200' elevation. (This area is accessible by rental car, but to go higher on the mountain, or to visit the astronomical facilities, one must have a four-wheeled drive vehicle!) *Stenogyne* can be found growing intertwined in the branches of another Hawaiian native, the yellow-flowered shrubby legume *māmane*, about which we will learn more later. This species of *Stenogyne* has abruptly right-angled stems with purplish-brown flowers and small leaves (hence, *microphylla*). The third species of *Stenogyne* featured here is *S. purpurea*, characterized by small, purple-pink flowers (Plate 61). This species occurs in wetter sites on Kaua'i where it can be seen growing beside the board walk heading towards the Alakai Swamp Trail.

      The authors of the *Manual* suggest that the ancestor of *Phyllostegia* may have been the original colonist from which the three genera of native mints arose. This kind of speculation is usually based on the species of interest exhibiting a suite of morphological features generally considered to be primitive (or less specialized) within the groups under study. In order to test this idea, it would be necessary to subject the group to rigorous examination including detailed analyses of anatomical and morphological features, structure and number of chromosomes, breeding biology and, ideally, sequence data for one or more genes. By comparing the outcomes of these analyses with available information on likely related genera, a better idea of evolutionary relationships within the group may be obtained. One problem with this particular group of genera in the Hawaiian Islands would be scarcity of plants belonging to *Haplostachys* (most are extinct). Because current DNA sequencing techniques can function with as little as a tenth of a

Plate 61. *Stenogyne purpurea*, a small flowered member of the genus that rewards those who venture into wetter spots on the islands such as along the Alakai Swamp Trail.

gram of plant material, data could likely be acquired for the one extant species, but patterns of structural variation for the extinct species would have to rely on available dried specimens, and chromosomal features would only be available had they been observed by earlier workers, an unlikely possibility.

An important factor in thinking about the evolutionary history of these genera involves the absence of oil glands, a structural feature often considered characteristic of most members of the mint family (Lamiaceae or Labiatae). The family is quite large, however, and consists of a number of groups that do not possess oil glands, at least of the sort and number normally associated with the 'true' mints. Therefore, there are two explanations for the absence of glands in Hawaiian mints: (1) the original colonizer did **not** possess glands; or (2) the original colonizer did possess oil glands, but over time lost them with the result being that all of its descendants also lack them. In order to support the first hypothesis, it would be necessary to find a genus of mints somewhere in the world that lacked glands and show that it is the closest ancestor to the Hawaiian plants. Gene sequence comparisons are very helpful in sorting out relationships of that sort. However, if the most likely ancestor of the island mints possessed oil glands (hypothesis 2), then we would have to account for the loss of this feature. The broader issues surrounding loss of antiherbivore protective chemistry has been discussed, in the absence of hard data in most cases, by several workers (e.g., Carlquist, 1980; Bohm, 1998). It is useful to review some of the ideas surrounding the possible function of glands and the oils they produce before we look at recent data that both help clarify the picture and illuminate a major problem in evolutionary problems of this sort.

Since plants cannot escape their enemies, they must resort to other methods to protect themselves. Several strategies have evolved to deal with threats from attackers whether they be microbes, insects, or higher animals. A simple solution is to lay down a thick cuticle on leaves and stems to provide some protection against teeth. An array of thorns also serves nicely to keep all but the most motivated at a safe distance. A fiendishly clever mutualistic association has been established between certain African acacias and species of ants. The tree provides a residence for the ants, and the ants pay their rent, so to speak, by attacking potential herbivores with their fierce biting attacks. Plants are also very skilled in the art of chemical warfare. One of the simplest chemical defenses seems to be the production of compounds that yield hydrogen cyanide (HCN) when plant tissue is disrupted. It has been speculated that HCN can work at two different levels. It may be toxic to microorganisms by interfering with their metabolism in one way or another, or the compounds that yield HCN (cyanogenic glycosides is the chemical term) may deter herbivores simply because they are quite bitter. In order to be effective, a feeding deterrent does not have to kill an animal, it need only stop it from taking a second bite. Animals learn from unpleasant experiences, quite quickly in most cases, whereas being killed offers little in the way of educational benefit (as an animal behaviorist friend of mine puts it, other animals do not stand around taking notes).

It is often argued that organisms that move to islands will lose the defensive mechanisms that helped them survive in their continental environment. There has been a lot written about this subject (Bohm, 1998), but very few experimental studies have been published. Empirical evidence suggests that most groups of plants maintain the capacity to accumulate potentially deterrent chemicals when they migrate to islands, but quantitative, statistically-based studies are limited. One study that did address the question dealt with the occurrence of cyanogenic glycosides in plants on the Galapagos Islands compared to their nearest relatives on the South American mainland (Adersen et al., 1988). The study revealed a small but statistically significant reduction in the fraction of HCN-positive species on the islands relative to closely related continental species. In a related study, although based on a much smaller sample, there was evidence that certain defensive strategies of plants on the Channel Islands of California are somewhat reduced compared to related plants on the mainland (Bowen and Van Vuren, 1997).

With those background comments, we can now return to our discussion of the possible role of mint oils and why the capacity to make them might have disappeared from island species. In low concentrations, such as one might encounter in a candy or in cooking a spaghetti sauce, the amounts of actual oil chemicals is quite low. At higher concentrations, however, such as might be experienced by chewing mint leaves, the mixture can be quite bitter. It seems reasonable that these plants would be bitter to other herbivores as well (animal taste sensitivities are similar to those of humans). So, does the absence of oil glands in Hawaiian mints reflect a loss of that feature because the founding colonists were no longer under pressure to defend themselves against some herbivore in their new home? If this were the case, evolutionary selection would no longer favor individuals with oil glands. In time, the trait would disappear altogether and all further generations would be gland-free. One can do little more than speculate on the question without knowing the evolutionary history of the plants in question. Insular endemics provide a near perfect opportunity to study evolution of systems of this sort.

An answer to that question was supplied recently by Charlotte Lindqvist and V. A. Albert (2002), of the Natural History Museum in Oslo, Norway. These workers studied sequences of DNA from a large selection of mint genera looking for one that might have been the ancestor of the Hawaiian group. They reported three important findings: (1) the three native Hawaiian genera emerged from the study as a cohesive group (monophyletic), strongly suggesting that they are all derived from a single introduction; (2) the colonization occurred within the period 2.6 and 7.4 million years before the present, well within the time when at least one and likely several of the current high islands existed; and (3) the group is most closely related to the widespread genus *Stachys*, but specifically to members of this genus from the California Floristic Province. While this would appear to put the problem to rest, we have to look at *Stachys* in general and the Californian species of *Stachys* specifically to learn about their glands and their capacity to make the volatile chemicals we associate familiarly with the mints. In the first place, *Stachys* does not belong to the subgroup within the mint family that is characterized by those compounds. The lack of these compounds in the Hawaiian mints then does **not** represent a loss; the colonist lacked that feature to start with. Although many members of the mint family do not make the familiar array of volatile chemicals, they nonetheless do make a variety of other compounds that might serve equally well in providing an unpalatable nibble for potential herbivores. There is some information on the chemical constituents of some *Stachys* species in the literature but, unfortunately, we do not have any information on the chemical constituents of Hawaiian mints or possible ancestors from the continent that we could use to make comparisons. This is a very common problem.

Pittosporum

Pittosporum (Pittosporaceae) is a genus of about 150 species distributed from sub-tropical Africa through Asia to Australia, New Zealand, islands of the South Pacific, and the Hawaiian archipelago. Species endemism is high on many of the islands, e.g., as many as 50 on New Caledonia, 26 in New Zealand, and 10 in the Hawaiian Islands (plus two naturalized species). Of the 10 native species in the Hawaiian Islands seven are known from one island only. Plate (62) shows the Kaua'i native *Pittosporum gayanum* photographed along the boardwalk trail to the Alakai Swamp. It is easy to recognize with its shiny, bright green mature leaves and brown, woolly juvenile ones. Flowers of this species are small and borne on the stems. Fruits are brownish and woody as seen in Plate (63). The style is persistent and can be seen protruding from the top of the fruit. Hawaiians called this species *hō'awa* or *hā'awa*. C. E. C. Gemmill, of Waikato University in New Zealand, and colleagues (2002) examined the evolutionary relationships among the Hawaiian species and species from Australia, New Zealand, New Caledonia, Fiji, Rarotonga, Tahiti, and Tonga using DNA analysis. The natives exhibited 0.0% divergence indicating a single, fairly recent colonization with the closest related species (sister species) from Tonga and Fiji. Distribution of seeds of *Pittosporum* by birds has been suggested based upon the production by these plants of resin-coated, thus, sticky seeds.

Left:
Plate 62. *Pittosporum gayanum* with it's characteristic tan, furry juvenile leaves.

Above:
Plate 63. Fruits of *Pittosporum gayanum* which contain resinous seeds lending themselves to transport by birds.

Sanicula

Sanicula, commonly called snakeroot, belongs to Apiaceae (Umbelliferae), a family that is home to many familiar plants including angelica (*Angelica*), carrots (*Daucus*), celery and celariac (*Apium*), coriander (*Coriandrum*), dill (*Anethum*), parsley (*Petroselinum*), and parsnips (*Pastinaca*), among the edible members, and Queen Anne's lace (*Anthriscus*) among the inedible ones. The *Manual* lists 13 genera of plants from this family that occur in the islands, three of which contribute species to the list of natives, *Peucedanum*, *Sanicula*, and *Spermolepis*. *Peucedanum* consists of at least 100 mostly Eurasian species. The Hawaiian member of the genus is *P. sandwicense*, a cliff dweller, a photograph which was taken in the Limahuli Garden appears in Plate (64). *Spermolepis hawaiiensis* is one of five species in the genus; three others occur in North America and one in Argentina. Formerly known from a number of islands, the only currently known population in Nature was reported from O'ahu in 1988, which discovery removed it from the list of extinct species (temporary reprieve?). *Sanicula* consists of at least three dozen species many of which occur naturally in California, with one known from Baja California, and two disjunct in South America. Of the four species of *Sanicula* that occur on the Hawaiian Islands, three are considered endangered; only *S. sandwicensis* (Plate 65), which is limited in occurrence to East Maui and Hawai'i, seems safe.

A recent study of *Sanicula* by Pablo Vargas, now at the Royal Botanic Garden in Madrid, and his co-workers at the University of California at Berkeley (1998, 1999) using DNA sequence comparisons revealed that the Hawaiian species are descended from species that occur within the California Floristic Province. Dispersal by birds seems likely in view of the existence of hooked prickles on fruits of the Californian species. By comparing DNA sequences, those workers concluded that the mainland and island species shared a common ancestor 900,000 plus or minus 400,000 years

Plate 64. *Peucedanum sandwicense*, a cliff-dwelling member of the carrot family, here photographed in the Limahuli Gardens.

Plate 65. *Sanicula sandwicensis,* another member of the carrot family endemic to the Hawaiian Islands and a species closely related to a Californian native shown in the next plate. Photo by Bruce Baldwin.

ago. Taking the maximum estimate from this range, 1,300,000 years, they speculated that suitable habitats for colonization could have existed on Kaua'i, O'ahu, and West Maui. Plate (66) features "footsteps of spring," *Sanicula arctopoides,* a mainland species thought to be a close relative of *S. sandwicensis.*

Silene, the campions

Silene, known to most gardeners as "campion" or "catchfly," belongs to Caryophyllaceae, a moderately large family with at least 2,000 species. The family is also home to such familiar plants as the showy *Dianthus* (pinks and carnations) and the weedy *Stellaria.* Eight species of *Silene,* seven of which are native, occur on the Hawaiian Islands. The native species can readily be sorted into two groups, one of five

Plate 66. *Sanicula arctopoides,* a close Californian relative of the Hawaiian members of the genus. Photo by Bruce Baldwin.

Plate 67. *Silene hawaiiensis,* a Hawaiian campion that requires comparatively young lava for
growth. Photo by Anna Westerbergh.

Plate 68. *Silene struthioides,* another campion
restricted to young lava sites. Photo by Anna
Westerbergh.

and the other of two species, which the authors of the *Manual* suggest may be the
result of two independent colonizations. Information is not available to address that
problem, but a study of the smaller group has revealed interesting relationships
between them and the habitats in which they live. The two, both of which are pioneer species, are *S. hawaiiensis* and *S. struthioides,* Plate (67) and Plate (68), respectively. *Silene hawaiiensis* occurs on younger soils (the *Manual* refers to them as decom-
posed lava and ash) on southern volcanoes on the island of Hawai'i, while *S.
struthioides* occurs only on Haleakalā on East Maui, and on Mauna Kea on Hawai'i.
A study of this species pair by two Swedish workers, Anna Westerbergh and Anssi
Saura (1994) revealed that levels of genetic variation in populations on the younger
island are extremely low, and that the plants show almost no morphological variation as well. By contrast, populations of *S. struthioides* on Maui and on Mauna Kea
are somewhat more variable. This is typical of progenitor:daughter species pairs
where the daughter species carries only a small sample of the variation exhibited by
the ancestor. This can readily be appreciated when one considers that colonization
is a chancy business and that only an occasional seed may make the successful journey to a new site. The available data do not allow us to say how long each of these
colonization cycles takes, but sites on Haleakalā cannot be older than about three
quarters of a million years. The source of seeds from which the Haleakalā populations
arose could have been from plants on West Maui, which is older than East Maui, or
possibly Moloka'i, which is somewhat older still. Suitable sites for these species no
longer exist on either of those islands. Whatever the past was for these species, their
future is dependent on the availability of new lava. It is possible that colonization of
lava fields currently being deposited on the eastern and southeastern shores of
Hawai'i may occur, and that eventual emergence of the next volcano (Iniki appears
to be on its way!) will offer a secure future for these species.

Tetramolopium

In several places in this book we will meet the suggestion that some, perhaps most, members of the Hawaiian native flora have affinities with species native to the southwestern Pacific islands and bordering mainland. For example, we have seen that *Metrosideros polymorpha*, 'ōhi'a , has been shown to be related to New Zealand ancestors who appear to have made it to the Hawaiian Islands by way of the Marquesas. Recent studies of the genus *Tetramolopium*, a member of the sunflower family with small, daisy-like flower heads, have also revealed close relationship with species from southwestern Pacific, particularly those that are native to New Guinea, where 25 of the 36 or so known species occur. Ten other species are native to the Hawaiian Islands; the eleventh occurs on Moloka'i as well as on Mitiaro in the

Plate 69. *Tetramolopium humile* is a high elevation native on the islands; most closely related to species from high elevation sites in New Guinea.

Cook Islands (ca. 20°S, 160°W). The status of most species of *Tetramolopium* on the Hawaiian Islands ranges from rare to extinct; four are endangered. The only species that one can reasonably expect to find in the wild without serious effort is *T. humile* (Plate 69). This small, crevice-dwelling species can be found in the summit area of Haleakalā and in exposed, dry sites on Mauna Loa, Mauna Kea, Hualālai, and Kīlauea on the Big Island. The Haleakalā area is a good deal easier to visit than most of the other sites. Plate (1) shows a typical habitat which *T. humile* is sharing with a species of *Hypochaeris*, a common lawn weed. This situation is not uncommon and will be addressed further in Chapter Three.

Studies of morphology, anatomy, chromosomal features, and protein electrophoresis of all Hawaiian *Tetramolopium* species by Tim Lowrey of the University of New Mexico and others revealed that the genus comprises a natural evolutionary group, that is, they are all descended from a single colonizer (Lowrey, 1986, 1995). Using the protein data it was possible to identify *T. humile* as the Hawaiian species most resembling New Guinea ancestors. This is of particular significance because the habitat in which *T. humile* occurs, the high mountains on Maui and Hawaiʻi, is typical of the tropical alpine habitat occupied by the genus in New Guinea. Carlquist (1980) suggested that an important requirement for successful colonization was the availability of habitats similar to those in which the colonizing species grew. The transfer from alpine habitat in New Guinea to alpine habitat in the Hawaiian Islands satisfies this requirement. In addition to pointing to a likely ancestral type, the protein data allowed an estimate to be made as to when the colonization might have occurred, and that in turn suggested where the initial "landing" might have been. Genetic divergence between the most ancestor-like species, *T. humile*, and the most advanced, or specialized, species on the islands, *T. lepidotum*, suggested a time of divergence between 600,000 and 700,000 years before the present. Where were likely habitats at that time? It has been estimated that Haleakalā on East Maui had reached an elevation of over 11,000' by 400,000 years before the present (it is a bit over 10,000' now). At that elevation, suitable habitats would have been available for colonists. And soon (in the geological time scale) emergence of the high mountains on Hawaiʻi would have afforded habitats to which *Tetramolopium* was already accustomed. Dispersal to other islands may not have presented much of a problem, at least with regard to over-water transport, since Maui, Molokaʻi, and Lānaʻi were all part of Maui Nui at the time. An interesting situation exists with regard to *T. rockii*, which is native to lithified dunes on the northwestern shores of Molokaʻi. Studies of the geology of these dunes have shown them to be about 15,000 years old, requiring *T. rockii* to be a species of very recent origin. Finally, the absence of *Tetramolopium* species on Kauaʻi suggests that that island had already aged to the point where high, or even medium, elevation habitats were no longer available.

The origin of the Cook Island *Tetramolopium* remains to be addressed. Three possibilities exist to account for the current distribution of the genus, assuming that it originated in New Guinea, as is generally accepted. The simplest explanation would be independent colonizations of the Cook Islands and the Hawaiian Islands. The second would require an early colonization of the Cook Islands with subsequent migration to and diversification of the genus on the Hawaiian Islands. In the third scenario, colonization of the Hawaiian Islands would be followed by diversification and subsequent colonization of the Cook Islands from a source on one of the Hawaiian Islands. Again, the protein data provide a resolution to the problem. Those data showed that the Cook Island plants appear to be most closely related to *T. sylvae* from Molokaʻi. Thus, colonization of the Hawaiian Islands followed by diversification and long distance dispersal of a propagule to provide the Cook Islands with their representative is the currently accepted view of the evolutionary history of the genus.

Vaccinium—'ōhelo—the blueberries

Blueberries belong to *Vaccinium*, a widespread genus having as many as 450 species worldwide. *Vaccinium* is a member of Ericaceae along with such other familiar genera as *Arctostaphylos* (bear berry), *Gaultheria* (wintergreen), *Erica* (the heathers), and *Rhododendron* (rhododendrons and azaleas). In addition to the various species of blueberries that grow wild in northern temperate zones, there are several species, including the cranberry, that are grown commercially. According to the *Manual*, there are three species of blueberries native to the Hawaiian Islands, although there is some doubt that recognizing only three can adequately accommodate the considerable levels of variation seen in these plants in the wild. The Hawaiian species, along with two that are native to the Society Islands, were thought to comprise a closely related group within the genus (but see below).

To give the reader some idea of the efforts to accommodate the variation that seems to characterize the Hawaiian species, it is interesting to note that under *Vaccinium calycinum* there are listed 17 synonyms, under *V. dentatum* 11, and under *V. reticulatum* nine. These synonyms represent botanists' attempts over the years to come to grips with what they were observing in the field. In an effort to eliminate, or at least reduce, environmental effects "common garden" experiments can be done. Seeds from as many of the variants as possible are germinated and then the seedlings grown under uniform laboratory or greenhouse conditions. Differences that are maintained under these conditions are considered to be under genetic control and can then be translated into botanical names that reflect the constant features. In the case of the Hawaiian blueberries, common garden studies revealed the existence of the three species named above.

Two of the native blueberry species are particularly easy to find in Nature. Both *V. calycinum* and *V. reticulatum* occur in abundance in Hawai'i Volcanoes National Park. *Vaccinium calycinum* (Plate 70) occurs beside the trail that runs along the western rim of Kīlauea Iki. It is a shrub that can attain a height of a meter and a half, has pale green leaves, and large, red berries. Continuing along the trail one eventually descends

Plate 70. *Vaccinium calycinum*, the largest of the Hawaiian blueberries.

Plate 71. *Vaccinium reticulatum*, 'ōhelo, photographed near Kīlauea Iki on Hawai'i.

Plate 72. *Vaccinium reticulatum* from a site on Haleakalā on Maui.

into the crater itself where the trail leads across the hardened lava, up the northern face of the crater, and back to the parking lot. *Vaccinium reticulatum* occurs abundantly on the crater floor where its often bright red terminal leaves, red flowers, and bunches of berries stand out brilliantly against the black lava. This species can also be found along the Devastation Trail (mentioned above), which leads to the eastern rim of the crater, and almost anywhere else in the Park in suitable habitats. Plate (71) features *V. reticulatum* photographed in the Devastation Trail area, while Plate (72) shows the same species growing along the

Plate 73. *Vaccinium dentatum* from a site on Kaua'i.

Halemau'u Trail on the northern flank of Haleakalā. The third species, *V. dentatum*, (Plate 73), occurs in wetter areas near bogs and can be found, for instance, in the vicinity of the 1 mile marker along the Pihea Trail on Kaua'i.

In the Hawaiian language, both *V. reticulatum* and *V. dentatum* are called "'ōhelo." *Vaccinium calycinum* is called 'ōhelo kau la'au, which literally means 'ōhelo placed on trees, recognizing its shrub-like habit in contrast to the usually smaller plants of the other two species. The early Hawaiians were well aware of the choice berries produced by these species (the tree form was considered inferior to the other two), with uses that continue to this day. Harvested in the Autumn, the berries are used to make very tasty jellies and jams. The plant is considered sacred to the volcano goddess Pele, to whom offerings of branches were made in the past—and continue to be—by throwing them into the caldera of Kīlauea Volcano. Air-dried leaves of ʻōhelo are also used to make *lei*.

Relationships within the genus *Vaccinium* have long been the subject of debate. The three Hawaiian species and a species from southern Polynesia (*V. cereum*) have generally been thought to comprise a group separate from others. Recent studies by E. A. Powell and K. A. Kron of Wake Forest University (2002) using DNA data have supported the idea that the three Hawaiian species form a tightly knit (monophyletic) set. *Vaccinium cereum* , however, appears to be the product of hybridization, possibly between Asian species, and that its similarity to the Hawaiian species is more apparent than real. The results also suggest that the Hawaiian species belong more comfortably within a group of blueberry species many members of which occur in Pacific Rim countries ranging from Japan to Mexico. That the Hawaiian species are the result of a single colonization seems certain, but we still do not know who the nearest relative is.

In contrast to the high degree of structural and flower pigment variation that seem to characterize these species, analysis of tannins and related compounds in leaves of all three species, collected from a wide selection of habitats and from several islands, revealed only minor quantitative differences (Bohm and Koupai-Abyazani, 1994). The profile of compounds that we identified are of the same general sort seen in North American species, with whom they are thought to be related. Tannic materials are generally considered to be effective feeding deterrents because of their bitterness and/or astringency. In contrast to the apparent loss of oil or production of other deterrent compounds in the mints, Hawaiian blueberries have retained their capacity to produce chemical substances that can serve as antiherbivore agents.

Viola—Hawaiian violets

This example is placed near the *Tetramolopium* story in order to emphasize the other extreme in the range of possible sources of propagules from which the island flora has arisen: *Tetramolopium* from the southwestern Pacific, *Viola* from the north. The common violet belongs to the genus *Viola* and is one of about 20 genera in Violaceae. (Don't confuse the true violet with the African violet, which belongs to the genus *Saintpaulia* of the Gesneriaceae, a totally different group of plants.) Violaceae is represented in the Hawaiian Islands by two genera, the native *Isodendrion* with three rare and one extinct species, and *Viola* with seven native species. All of the Hawaiian violets have woody stems which led earlier botanists to suggest that the island species represent an ancient lineage (woodiness is generally an ancestral condition and most other members of Violaceae are non-woody).

An examination of the Hawaiian members of *Viola* was undertaken by H. E. Ballard of Ohio University and K. J. Sytsma of the University of Wisconsin (2000) with a view to answering five questions: (1) Have all Hawaiian violets arisen from a single colonization? (2) Do they represent an ancestral lineage within the genus as indicated by their woodiness? (3) Are they derived from a group of Central American and Andean violets as suggested by earlier workers? (4) Is their woodiness really an ancestral feature or could it be secondarily derived as has happened with several other, unrelated groups of island plants? (5) When did colonization occur? Selected DNA sequences were chosen as a possible source of answers to these questions. All of the Hawaiian species were sampled as were an additional 40 species representing the wider range of *Viola* including those from North, Central, and South America, Asia, and Europe.

Analysis of the DNA results provided answers to all of the questions. They clearly showed that Hawaiian violets are the result of a single colonization that occurred no more than 3.7 million years ago with colonization first occurring on Kaua'i with subsequent radiation to O'ahu. Analysis also placed the Hawaiian species among the most advanced species in the genus from which result it is possible to conclude that woodiness is a derived feature, that is, a feature that has aided in the group surviving on the islands. Most analyses of DNA data point to a group of species within a genus as the most likely source of an ancestor. In the present instance, the data pointed to a single species as most closely related to the island taxa, namely *Viola langsdorffii* from the Arctic. It was possible to narrow the source of colonizer even further when chromosomal information was taken into account. *Viola langsdorffii* consists of several different chromosomal races throughout its range. Plants from Kamchatka have been shown to be hexaploids (six sets of chromosomes) with n = ca. 30, while those from Japan are decaploids (ten sets) with n = ca. 50. Plants from the Queen Charlotte Islands, British Columbia are known to have twelve sets of chromosomes (dodecaploids) with n = ca. 60. Although chromosome numbers have not been determined for the Alaskan Arctic plants, it has been speculated that there should be populations with the "missing" octaploid (eight sets) number of n = ca. 40. It is, thus, very interesting to note that chromosome numbers for Hawaiian species of *Viola* have been recorded as n = 38, 40, and 41-43.

A powerful piece of information emerged from the analysis of the DNA sequences which establishes without any doubt that the Hawaiian violets constitute a monophyletic group, that is, a group of species that have been derived from a single ancestor. All species share a 26-base pair deletion in the DNA sequence used in that work (it's called the ITS sequence). Alterations of a DNA sequence of this magnitude would not be expected to occur randomly in unrelated species. Their presence in a group of species is an extremely strong indication of close relationship.

It is not unfair to stress that an Arctic origin of the island violets came as a surprise. It is the first such example and involves a species, *V. langsdorffii*, that had never been considered as a possible ancestor of the island violets. The authors commented on the obvious bias of workers who usually search for tropical origins of Pacific island flora: one does not normally look for an Arctic origin for a tropical plant. They go on to ask if other propagules of northern origin have been delivered to the islands—they are regularly visited by migrating birds—and whether the propagules simply haven't found a compatible niche. In the case of the Hawaiian violets, several occur in cool, wet habitats at higher elevations, conditions that are not drastically different from those in their home range.

Wikstroemia

Wikstroemia (Thymelaeaceae) is a genus of some 50 species that occur in southeastern Asia, Australia, Fiji, and the Hawaiian Islands. The *Manual* lists 12 native species, including one that may be extinct, and suggests that they were all derived from a single introduction. Bo Peterson of the Botanical Museum at the University of Göteberg, writing in the *Manual*, points out the difficulty in dealing with patterns of variation within *Wikstroemia*, or more precisely, lack of clear-cut patterns of variation. Few characters, he notes, are consistently reliable as indicators of species limits. The lack of significant variation patterns may have been recognized by the Hawaiians who had a single word, 'ākia, for this entire group of plants. Just how many species of *Wikstroemia* exist on the islands also appears to be an open question. The Swedish botanist Skottsberg recognized 24 named species in a work published in 1972, as opposed to the much larger number noted above. An interesting attempt to resolve this situation was described by Stephanie Mayer, then at the University of California at Berkeley, who applied statistical methods to an analysis of 20 vegetative and inflorescence characters and 16 flower features. Numerical analysis (clustering) of values representing these variables failed to reveal any clear-cut groupings, which is in general agreement with the problems of identifying species using orthodox methods. Dr. Mayer put forward two hypotheses to account for these observations. The first suggested that the genus is in the process of diverging, but it is too soon to see significantly differentiated individuals (species). The second hypothesis suggested that the present variation is unstable and that distinctions between

Plate 74. *Wikstroemia uva-ursi.*

"species" will continue to be blurred because of free flow of genes among all members of the genus. It is interesting to contrast this system with the silver-swords where gene flow is highly restricted in Nature.

Wikstroemia provided a very tough fiber from which Hawaiians made twine and rope. Several preparations of these plants were used as a laxative and as a treatment for asthma (not the same preparation one presumes!). It has also been reputed that preparations of 'ākia, the generic Hawaiian term for the genus, were used as a "terminal punishment" for certain wrongdoers. Examination of the chemical literature for Thymelaeaceae reveals several groups of compounds, alkaloids among others, that could easily provide that means of exit. Preparations have also been used for stupefying fish, which is in line with usage of other members of family as fish poisons in India.

Wikstroemia uva-ursi, the species illustrated in Plate (74), shows the characteristic small, yellow flowers of the genus. This species is frequently used as a border shrub around hotel buildings and along walkways.

This following set of examples includes several genera with native species that have not been studied using macromolecular techniques. The examples are presented alphabetically using common names, e.g., rose family, sandalwood family, genus, or family names with modifiers, e.g., Rubiaceae—the coffee family. We'll begin with an example from the amaranth family.

Amaranths

Amaranthaceae is the family that is home to the common garden plants amaranth, (*Amaranthus*), celosia (*Celosia*), and globe amaranth (*Gomphrena*). *Amaranthus albus* is one of the so called pig-weeds many of which are naturalized pests in North America where they occur in dry, waste areas. Amaranthaceae are represented in the Hawaiian Islands by six genera two of which are native or nearly so, and two others that are more widely distributed but represented on the islands by native species. Six species of *Amaranthus* occur on the islands one of which, *A. brownii*, is native but rare, surviving only on Nihoa where a dozen plants were counted in 1983. (Nihoa lies to the northwest of Kaua'i, see Chapter One). The others are widespread, weedy species. I have not visited Nihoa so I have not seen this species of *Amaranthus*, nor have I seen any of the three native species of *Achyranthes*, not surprising considering that one of them is extinct and the other two are listed as endangered and are difficult to find.

Charpentiera, known to the Hawaiians as *pāpala*, consists of six species five of which are native to the Hawaiian Islands; the sixth occurs on the Cook and Tubuai Islands. The Tubuai Islands, which lie east of the Cook Islands and south of the Society

Plate 75. *Charpentiera elliptica,* native to and quite common on Kaua'i.

Islands, are referred to as the Austral Islands in the *Manual*. Tubuai itself lies at 23°23'S, 149°27'W. The disjunction between the Hawaiian Islands and islands in the South Pacific is known for other Hawaiian taxa, having been encountered with *Tetramolopium* in the Cook Islands, and going a bit farther, with the Marquesas in reference to *Metrosideros*. Plate (75) shows the Kaua'i native *C. elliptica* photographed at the Limahuli Garden. The characteristic dark purple fruits of this plant are easily visible in the picture.

Certainly among the most unusual uses for a native plant involved the activity called *'oahi*, which relied upon the highly flammable pith of species of *Charpentiera* (and an offshore wind). Young men would climb to the top of certain cliffs on the Nā Pali coast of Kaua'i, collecting dry branches of *C. elliptica* along the way. At the top they would light them and throw them into the air, sending a shower of sparks from the burning pith out over the ocean and onto spectators who waited in canoes below. Apparently, to catch a burning ember and brand one's self served as a souvenir of the event (Krauss, 1993). Only a few promontories were used for these festivities, one of which, Makana (elev. 1280 ft), stands above Kē'ē Beach at the end of the road on Kaua'i's north shore.

Argemone glauca—pua kala, kala, naule, pōkalakala, prickly poppy

Argemone is a member of the Papaveraceae, the poppy family. Thirty species are known two of which occur in the Hawaiian Islands: the naturalized *A. mexicana,* which, as its name suggests, is home in Mexico (and the West Indies), and *A. glauca,* which is native to the Hawaiian Islands where it grows in moderately dry areas on all of the main islands. The plant illustrated in Plate (76) was found along the jeep trail on Mauna Kea at about 9,000' elevation. According to the Manual, two

Plate 76. *Argemone glauca* photographed along the Jeep Road on Mauna Kea.

varieties of *A. glauca* are recognized, var. *glauca* and var. *decipiens*. The pictured plant is var. *decipiens* whose range on the Island of Hawai'i includes dry sites on the saddle between Mauna Loa and Mauna Kea. Variety *glauca* is restricted to the South Point area of the island, which, incidentally, is the southernmost point in the United States.

Sap and seeds of this plant were used by Hawaiians as an analgesic and narcotic. This is not surprising in view of the many pharmaceutically important drugs that have been isolated from other members of the Papaveraceae, in particular, morphine and related alkaloids. Morphine, of course, continues to be the most effective naturally occurring analgesic known.

Several aspects of this plant are worth noting. In contrast to the vast majority of Hawaiian native plants which are perennials, *Argemone* is a biennial, requiring two years to complete its life cycle. It is not uncommon for plants that become transplanted to island habitats to go through a biennial life cycle phase before becoming perennial. The fact that *Argemone* retains the biennial life style suggests that it may be a relatively recent addition to the islands' flora. Both Carlquist (1980) and Kepler (1998) note that its arrival on the islands may have just predated arrival of the first Polynesians. Another feature of the plant suggests that it may well be comparatively new on the islands, namely that it has retained its extremely sharp armaments. There is evidence to suggest that a plant may lose some or all of its defensive structures—barbs, prickles, thorns—as a result of moving into a new habitat that lacks the threat against which the devices were needed in its original habitat. Carlquist (1980) suggests that had this species invaded the islands a million years ago, it may well have lost most of its defensive spines by now. The same argument has been advanced for the reduced prickliness of native Hawaiian raspberries compared to mainland North American species, among others. It has also been suggested by several authors that chemical defenses are also likely to be lost as the newcomer becomes accustomed to its new home. *Argemone* has obviously not lost its capacity to manufacture an array of narcotic compounds. This generalization has been challenged by the present author (Bohm, 1998) who holds that maintaining a certain level of defensive chemistry is in the general interest of plants. After all, moving to an island does not insure a pest-free existence. Most plant chemicals used for defense against herbivores are intensely bitter and work by tasting bad, although many of the compounds are toxic in the usual sense to microorganisms.

Astelia menziesiana—*kaluaha, pua'akuhinia*

Astelia menziesiana—the name honors Archibald Menzies—is one of three species of this genus of lilies that are native to the Hawaiian Islands. The genus itself is modest in size, perhaps as many as 25 species, ranging from Chilé in the east to Mauritius and Réunion in the west with species known from other Pacific islands. This species is characterized by an attractive head of orange berries as seen in Plate (77) while the flowers, yellow or greenish, are less conspicuous. The species has male and female flowers on separate plants. This species can be terrestrial or epiphytic and occurs in mesic to wet forests and bogs on all of the major islands. Carlquist (1980) talks about the fruits of Astelia, and other plants, being particularly attractive to birds who were likely the means by which ancestors of the island species arrived in the first place.

Plate 77. *Astelia menziesiana*, a member of the lily family showing off its colorful fruits.

Plate 78. *Broussaisia arguta*, the only species in this Hawaiian native genus; related to the hydrangea.

Plate 79. *Hydrangea macrophylla*, the common horticultural hydrangea.

Broussaisia

Broussaisia, a member of Hydrangeaceae—the hydrangea family, has the distinction of being one of the native genera in the Hawaiian Islands. The sole species, *B. arguta*, is a fairly common member of medium to moderately wet forests on all of the main islands. I have seen it in the forests on Hawaiʻi (beside the road from Hilo to Hawaiʻi Volcanoes National Park) and along the Pihea Trail on Kauaʻi. The species is characterized by a large pink to purple flower head, as can be seen in Plate (78). Male and female flowers occur on separate plants. The genus appears to be related to the common, blue-flowered horticultural *Hydrangea* with which it bears a striking resemblance (Plate 79).

Chamaesyce halemanui—ʻakoko, koko, ʻekoko, kōkōmālei

Chamaesyce is a genus of some 250 species widespread in the New World. The genus belongs to the Euphorbiaceae, the family to which several very well-known plants belong, among which are the common garden croton (*Croton* species), rubber (*Hevea brasiliensis*), castor bean (*Ricinus communis*), and, in the Hawaiian Islands, *kukui* (*Aleurites moluccana*), which we will meet below. The Hawaiian Islands are home to 22 species of *Chamaesyce* 15 of which are native. Our representative in this genus is illustrated in Plate (80). This plant was found in a hotel garden on Kauaʻi and was unlabelled. Following the description in the *Manual*, this specimen seems to be most closely akin to *C. halemanui*, a native species that is further restricted to Kaua'i.

One of the distinctive features of members of *Chamaesyce* is the presence of pairs of opposite leaves which is clearly visible in the plate. Another characteristic of many members of the Euphorbiaceae is the production of a milky sap, or latex, often in copious amounts—think of the rubber tree—which can be extremely bitter to taste. Species of *Chamaesyce* produce a milky sap which was used as a component of some paints (Krauss, 1993). The sap can be extremely bitter and as such can serve as a feeding deterrent to most herbivores: the first nibble can readily convince all but the most hungry beasts to look for better tasting greenery elsewhere.

In the discussion of *Argemone* above, it was noted that the plant's means of defense against herbivores—spiny leaves, and I added bitter chemicals—are still maintained because the plant (or its immediate ancestor) appears to have arrived on the islands only recently (pre-Polynesian arrival). I have seen no argument that *Chamaesyce* is a recent arrival, yet it has maintained a nasty array of unpleasant chemicals in its sap that are very effective in defending against herbivory. I submit this as another example of a plant that has adapted to a certain strategy and maintains it even in surroundings different from its ancestor's original home.

Plate 80. A species of *Chamaesyce* of the euphorb family.

Plate 81. *Chenopodium oahuense.*

Chenopodiaceae

Chenopodiaceae is a moderately large, cosmopolitan plant family with several very familiar members including edible species such as beets and Swiss chard (*Beta vulgaris*), spinach (*Spinacia oleracea*), and *Salicornia* (glasswort, marsh samphire). The weedy "Russian thistle," *Salsola kali*, also occurs in dry sites on Hawai'i but apparently has not yet become a major pest (as it has done in North America). The genus of interest here, *Chenopodium*, consists of perhaps 100 species and includes the South American grain crop quinoa (*C. quinoa*). Six species of *Chenopodium* are known in the Hawaiian Islands, only one of which, *C. oahuense*, 'āweoweo or 'āheahea, is native. This species bears a close resemblance to the well known weedy "lambs quarters," also sometimes called "pig weed," although the latter term frequently refers to other, but related species. The range of *C. oahuense* includes all of the main islands, except Kaho'olawe, and extends as far northwestward as Lisianski Island.

Although the *Manual* describes *C. oahuense* as "scentless to weakly scented shrubs," the plants that I encountered on Mauna Kea near the Hale Pōhaku Visitor's Center had a distinctly unpleasant, ammoniacal aroma. It is entirely possible that dried specimens lose their odor after time, but in the field, the plant is very easy to identify by nose! Plate (81) shows a typical array of the leaves of this species, which are pale green in color, soft to the touch, and smelly. A noteworthy feature of *C.*

oahuense is that it is the only member of the genus to attain the status of tree, although it rarely ever reaches two meters in height. Even in its shrubby form, it is the woodiest species of *Chenopodium* in the world (Carlquist, 1980). Hawaiians used wood of this plant, and other species, to fashion the curved shafts of fishhooks and ate the cooked leaves in times of scarcity (Krauss, 1993).

Cyrtandra—haʻiwale, kanawao keʻokeʻo

With 53 species listed in the *Manual*, *Cyrtandra* is the largest genus in the Hawaiian flora. The genus is one of 139 that make up the Gesneriaceae, which accommodates about 2,900 species in all (Mabberley, 1997). Horticulturally, the most commonly met member of this group is "gloxinia," which, unfortunately, belongs to the genus *Sinningia* (*S. speciosa*). True *Gloxinia* is a genus in its own right (although the two are closely related). Hawaiian members of *Cyrtandra*, all of which are native, are extremely variable and often occur within quite limited geographical areas. An excellent place to see a species of this genus is in the area surrounding the Thurston Lava Tube where *C. platyphylla*, ʻilihia, occurs in shaded sites. This species is characterized by large, soft, woolly leaves and white to pale green flowers (Plate 82). I've always felt that a plant with leaves this size deserves larger flowers.

Plate 82. *Cyrtandra platyphylla*, characterized by round, woolly leaves and small flowers.

Geranium

Before we look at the native geraniums of the Hawaiian Islands, it is necessary to comment on the name geranium itself. Some common horticultural geraniums may actually belong to the genus *Pelargonium*, which also belongs to the family Geraniaceae. *Pelargonium* is a genus of about 280 mostly African species, but with representatives in Saudi Arabia, St. Helena, Tristan da Cunha, southern India, Australia, and New Zealand. *Geranium* proper consists of about 300 species mostly from montane tropics but with 38 species in Europe. There are also several species native to Malesia and the Hawaiian Islands. Members of this last group are the subject of this section. Geraniaceae is represented on the Hawaiian Islands by two gen-

Plate 83. *Geranium cuneatum* ssp. *tridens* is native to the slopes of Haleakalā on Maui while the other three subspecies occur only on Hawaiʻi.

era, *Erodium* and *Geranium*. The former, commonly called stork's bill, is represented on the islands by one naturalized species, while 11 species of *Geranium* have been reported. Six of the *Geranium* species are native to the islands and have a suite of characteristics that have prompted some botanists to recognize them as a genus in their own right (*Neurophyllodes*). Authors of the *Manual* maintain the group as species of *Geranium*, however, and we will follow that convention here. Of the six native species, two are listed as rare and two are considered endangered. One of the endangered species is G. *arboreum*, a red-flowered, bird-pollinated species that is unique in the family and apparently restricted to a single gully at about 4,500' elevation on Haleakalā. My only attempt to locate this species was cut short by a torrential downpour the likes of which I hadn't seen since witnessing monsoon rains in India and Thailand.

Much more common, and considerably easier to find in the field, is G. *cuneatum*. It does help, though, if you know where a particular specimen came from in order to identify it to the level of subspecies, four of which are recognized. Three of the subspecies occur only on Hawaiʻi and can be distinguished by location (which volcanic mountain) and by elevation of the collection site. The fourth subspecies, G. *cuneatum* subsp. *tridens*, occurs only on Haleakalā and is easily distinguished from the others, obviously by geography, and by the number of teeth at the tips of the leaves. Its subspecific name, *tridens* refers to its characteristic three-toothed leaf tips. At least one botanist recognized this form as a species in its own right. This plant is

Plate 84. *Geranium cuneatum* ssp. *cuneatum,* one of the subspecies that grow on Hawai'i. This one is from the eastern slopes of Mauna Kea at about 9,200' elevation.

common along the Halemau'u trail on Haleakalā, but it can also be seen near the Visitor Center and along the Hosmer Grove Road. It is illustrated in Plate (83). *Geranium cuneatum* subspecies *cuneatum* is illustrated in Plate (84). This plant was found on Mauna Kea in the Wailuku River drainage at about 9,200' elevation. Note the larger number of teeth on the leaf tips.

A recent application of DNA sequencing techniques to all native Hawaiian species revealed that they are more closely related to each other than they are to any other group of *Geranium* species (Pax et al., 1997). The next closest relationship was with two species from Mexico, followed by a connection to a North American species. The least closely related were species from Asia and Australia. The suggested relationship with American species is in line with a postulated relationship with species from Peru made by the Hawaiian botanist F. R. Fosberg. In situations like this it is usual for workers to conclude that a larger number of species ought to be tested, and that additional gene sequences should be investigated in an effort to detect any diversity that may exist among the potentially closely related species.

Ginseng family—*Cheirodendron*

Araliaceae is home for several well known plants including ginseng (*Panax ginseng*) cultivated for its medicinal action, the decorative octopus tree (*Schefflera* species), English ivy (*Hedera helix*), and *Fatsia japonica* (often referred to in the horticultural trade simply as "japonica"). Anyone who has ever tangled with "devil's club" while

Plate 85. *Cheirodendron platyphyllum* ssp. *kauaiense* occurs only on Kaua'i while its sister subspecies is found only on O'ahu.

Plate 86. *Cheirodendron triphyllum* growing on Hawai'i.

hiking in forests in western North America will have encountered *Fatsia* (syn. *Oplopanax*) *horridus* (truth in advertising!).The family is a moderately large one with some 50 genera that include about 1,300 species. The Hawaiian Islands have a fair representation of the family including several interesting and attractive natives. English ivy and the octopus tree are naturalized species, whereas some or all of the species of the remaining four genera are native to the Hawaiian Islands.

Cheirodendron, for example, is represented in the islands by five species with the sixth native to the Marquesas. *Cheirodendron platyphyllum*, the 'ōlapa or lapalapa (Plate 85), is an attractive tree with leaves that flutter in the wind reminiscent of trembling aspen. I have seen

Plate 87. *Tetraplasandra waimeae,* one of six species of this native island genus.

somewhere that the Hawaiian name is onomatopoeic reflecting the sound made when the leaves slap together in the wind. Two subspecies are recognized, subsp. *kauaiense* which is native to Kaua'i, as its name implies, and subsp. *platyphyllum* which occurs on O'ahu. The Kaua'i native can easily be found along the Pihea Trail between the three-quarter mile and one mile markers.

Plate (86) features another native species in this genus, C. *triphyllum*. A useful feature in identifying this species is its central leaf which is frequently larger than the two on either side. This is a highly variable species with at least 15 varieties having been described by one authority. It occurs in mesic to wet forests on all islands, except Kaho'olawe. The photograph for the plate was taken at about 5,000' elevation along the Saddle Road on the Big Island.

One of the more spectacular native genera on the islands is *Tetraplasandra* with six species. *Tetraplasandra waimeae* is pictured in Plate (87). This specimen was encountered in the dense forest beside the Awa'awapuhi Trail. According to the *Manual,* this species is closely related to *T. waialealae* which is also native to the wet forests of Kaua'i. The authors of the *Manual* suggest that the two may be forms of a single species, but that additional study is necessary to determine if the differences that distinguish them are under genetic or environmental control.

Munroidendrum consists of the single species M. *racemosum* (a raceme is a kind of floral display). According to the Limahuli Garden guide book, this native of

Kaua'i is known from only four sites in Nature but does very well in cultivation. A mature tree and a younger specimen are on display at the Garden. A recent study strongly suggests that *Tetraplasandra*, *Munroidendron*, and *Reynoldsia* (a Hawaiian native genus with only a single species, *R. sandwicensis*) have evolved from a single colonial ancestor (Costello and Motley, 2001).

Korthalsella

You have to be sharp-eyed to spot members of this genus of mistletoe relatives. Species of *Korthalsella* (Viscaceae) are parasites that can be found growing on a number of Hawaiian trees. One of the Hawaiian words for this group of plants is *hulumoa*, which translates literally as "chicken feathers," in reference to the manner of branching of the plant. The genus consists of about two dozen or so species

Plate 88. This odd looking plant is a species of *Korthalsella*, a parasitic genus. Here it is living on a plant of *Vaccinium calycinum*.

ranging from Ethiopia, islands in the Indian Ocean, southeast Asia, islands in the southwestern Pacific, Australia, New Zealand, to the Hawaiian Islands. The largest concentration of species occurs in the Hawaiian Islands where one finds four native and two indigenous species. Plate (88) shows an example of a species of *Korthalsella* growing on *Vaccinium calycinum*. *Korthalsella* species are not specialists; it is possible to find them on most Hawaiian tree species.

Legumes

The next examples come from Leguminosae (or Fabaceae), commonly known as the pea family. This is the third largest flowering plant family with some 18,000 species widely distributed on all continents except Antarctica. The family is represented on the Hawaiian Islands by 114 species in 50 genera. Despite this large number of species, only 13 are natives and five of those are thought to be extinct. Four of the remaining species are indigenous and one was among the plants brought by the original Polynesian colonists; the remainder have been introduced to the islands by a variety of means, some of which were accidental. Three of the natives are important floristic elements in the islands, *koa*, *mamane*, and *wiliwili*.

Acacia koa-koa

The first of these, *Acacia koa*, is known simply as *koa*. *Koa* was a major element of higher elevation forests at one time where individual trees attained heights of 35 meters and diameters approaching three meters. *Koa* displays two growth forms. Juvenile plants have leaves that are finely divided, as shown in Plate (89) and light green. As the tree grows, the dissected leaves are replaced by mature, crescent-shaped

Plate 89. *Acacia koa*, showing the highly dissected juvenile leaves.

"leaves" that are darker in color and stiff as shown in Plate (90). These are not true leaves, but are actually flattened petioles or leaf stems called "phyllodes." Phyllodes perform the functions normally associated with leaves, i.e., photosynthesis and transpiration.

How *koa*, or an ancestor, might have reached the Hawaiian Islands was discussed by Carlquist (1980) who pointed out that the most closely related species, based on structural similari-

Plate 90. *Acacia koa*, showing the mature "leaves," which are technically phyllodes.

Native Plants and Flowers of Hawai'i

13 Different Postcards

Native Plants and Flowers of Hawai'i – The Hawaiian Islands, isolated by thousands of miles of ocean for millions of years, developed a unique array of native flowering plants. Some of the plants shown here rank among the rarest in the world, while others are still relatively easy to find. All of them are unique to Hawai'i. Human introduction of exotic species of plants and animals has resulted in the rapid extinction of many native plants in Hawai'i. Many others are now rare and endangered.

Printed in Hawai'i on recycled paper with soy inks by Valenti Print Group, Honolulu.

NPF 00

ties and tetraploid chromosome number, is thought to be A. *heterophylla* from Mauritius. The distance between Mauritius, which lies east of Madagascar in the Indian Ocean, and the Hawaiian Islands would seem to be a daunting barrier. We were reminded in his discussion, however, that only a single seed would have been needed to establish a new colony. The situation may be somewhat simpler, however. The authors of the *Manual* tell us that both species are related to an *Acacia* from Australia (A. *melanoxylon*). Although the distances are still considerable, it is probably easier to imagine seeds from an Australian ancestor giving rise independently to both the Mauritian and the Hawaiian species. It is likely that a detailed study of DNA divergence could shed light on this problem.

   Koa forests were common at higher elevations, up to about 6,000', on all of the major islands (except Kahoʻolawe and Niʻihau) in the past but heavy logging practices have resulted in severe reductions. An easily accessible *koa* forest can be found near the upper reaches of the Mauna Loa Road. The stunning effects of volcanic activity can also be appreciated as one drives through the Keʻamoku lava flow. Several *koa* trees can also be seen at the trailhead at Kīpuka Puaulu, which is near the lower end of the Mauna Loa Road. We will meet *koa haole* (literally, foreign koa), a member of the legume genus *Leucaena*, in the chapter on alien species.

   Koa was an extremely important plant in the life of the Hawaiian people. A tall, straight tree provided the master canoe builder with the raw material from which he could construct the vital means of transportation upon which the communities depended. The choicest trees came from forests near Hana, East Maui, and from the Hilo and Kona districts on Hawaiʻi. Cut, hollowed, and shaped, the *koa* trunk provided the hull of a sea-worthy craft. Kepler (1998) describes a double hull canoe from Maui that measured 120' (37m) long and 9' (3m) deep! These must have been truly magnificent specimens, indeed. Very few trees of this size have survived into modern times because of the critical importance of sturdy sea going crafts to the Hawaiians.

   On its own, however, the hull would not remain upright with its crew; an outrigger was required. The outrigger was usually made from the wood of *wiliwili* (*Erythrina sandwicensis*) and was attached to the hull by two struts made from wood of the *hau* tree, *Hibiscus tiliaceus* (more below). Gunwales of the canoes were fashioned from a yellow wood such as a *Bobea* species (Rubiaceae), but others, including ʻohiʻa, were used as well. Paddles were made of *koa* wood. Sails were made by plaiting leaves of the screwpine, species of *Pandanus*, known as *hala* or *pūhala*. Additional information concerning the ceremonial selection of the appropriate *koa* tree for canoe manufacture, not a simple matter by any means, and the procedures leading to the final product are described by Krauss (1993). Another sea-going craft, not as significant for survival as the canoe but nonetheless truly Hawaiian, was the surfboard.

Smaller boards were made out of slabs of *koa* wood, but for larger ones, where a lighter wood was needed, *wiliwili* wood was employed. Nonetheless, one of the larger boards made from wiliwili wood, described by Abbott (1992), was a bit under six meters in length and weighed 78 kg (174 pounds). Traditions fade, however; today's boards are made from fiber glass.

Sophora chrysophylla—māmane

The second major native legume on the islands is *Sophora chrysophylla*, known in Hawaiian as *māmane* (Plate 91). This species is characterized by bright yellow flowers, as indicated by the specific epithet. The seed pod, illustrated in Figure (5), has a very characteristic shape. Often, it is possible to find small, bright orange seeds of *māmane* lying on

Plate 91. *Sophora chrysophylla,* easily identified by its bright golden-yellow flowers.

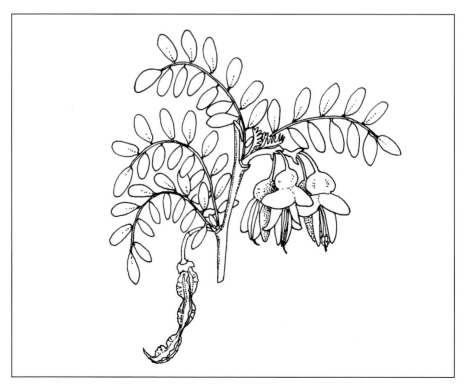

Fig. 5. *Sophora chrysophylla (māmane)* seed pods hanging on the lower left have a distinctive appearance.

the ground under the trees. *Māmane*, a major forest tree in subalpine areas on East Maui and on Hawai'i, can be conveniently seen, among many other places, on the Halemau'u Trail on Haleakalā, around the Visitor Center on Haleakalā, and in the forest near the Hale Pōhaku Visitor's Center on Mauna Kea. *Sophora* consists of about 50 species most of which occur in tropical and subtropical regions with only the one species present in the Hawaiian Islands. Carlquist (1980) suggested that *māmane* may be related to *Sophora* species from New Zealand. Other Hawaiian species may claim New Zealand as their ancestral home, although the path followed to get to their new home may have been somewhat roundabout, such as we saw in the case of '*ōhi'a* (*Metrosideros*).

Erythrina sandwicensis— wiliwili

The third legume mentioned above, *Erythrina sandwicensis*, or *wiliwili*, belongs to a genus of at least 100 species the majority of which are native to northern South America and islands in the Caribbean, with other centers of diversity in Africa and Asia. The Hawaiian species appears to be most closely related to the Tahitian species, *E. tahitensis*, and to a species from

Plate 92. *Erythrina sandwicensis* flowers. Photo by Ken Marr

northern South America, *E. velutina*. *Wiliwili* trees are native to dry forests from low to medium elevations on all of the main islands, but can be found now mostly in remnant patches. The tree is frequently planted for decoration. The seeds of *wiliwili* are used to make permanent *lei*. Plate (92) shows the flowers of *wiliwili*.

Lysimachia (primulas)

So far, we have met genera named after people (*Wilkesia*), a volcano Goddess (*Pelea*), and leaves that look like swords (*Argyroxiphium*), and species named after places (Sandwich Islands, Hawai'i, Tahiti, Mt. Waialeale), and people (David Douglas, King Kamehameha, Archibald Menzies), and various morphological features (*arborea* = tree-like, *latifolia* = wide leafed). In this section we will meet a species named after a meteorological event—in fact, a weather disaster of unprecedented severity. But first, we have to meet a new group of plants.

Plate 93. *Lysimachia daphnoides*. Photo by Ken Marr.

Lysimachia, which belongs to Primulaceae, the primrose family, enjoys a worldwide distribution although most of its species are native to the Himalayas and China. At least a dozen species occur in the Hawaiian Islands, all but one of which are native to the archipelago. The exception, *L. mauritiana*, occurs widely in the Pacific Basin, in eastern Asia, and, as its name suggests, Mauritius. At least one authority has suggested that the native species are sufficiently different from others in the genus to justify recognition as a separate genus, *Lysimachiopsis*. We await the results of DNA-based studies to see whether separate generic status is supported, as well as to gain some better idea of where the original colonizer(s) might have came from. An example of a native species is *Lysimachia daphnoides* (Plate 93) which occurs in the Alakai Swamp near Mt. Wai'ale'ale. One of the species in genus, *L. iniki*, came to our attention under unusual circumstances which I describe in the following.

I suspect most people think of the Hawaiian Islands as an idyllic holiday setting with warm temperatures moderated by the Trade Winds coupled with gentle, warm rains that support a lush tropical vegetation. Much of the time this view is a fair description of the islands. In late summer, however, when ocean temperatures rise above 80° in the eastern Pacific Ocean, there exists the threat of serious storms that can change the picture of the Hawaiian Islands dramatically. Storms can range in intensity from tropical depressions (maximum wind speed 38 mph), through tropical storms (maximum wind speed 73 mph), to hurricane (wind speed above 73 mph) (Juvik and Juvik, 1998). A brief look at the recent history of tropical storms affecting Kaua'i and O'ahu should give the reader an idea of the frequency of these events and their potential for devastation. In 1983, Tropical Storm Gil brushed the northern coasts of both islands inflicting damage, local but extensive in some cases, with 70 mph winds and associated high water. In August, 1959 Hurricane Dot crossed Kaua'i from the south-southwest causing an estimated six million dollars in damage. In November, 1982 Hurricane 'Iwa struck Kaua'i from the southwest wreaking damage estimated at two hundred and thirty-nine million dollars. But the worst was yet to come.

It so happened that the American Botanical Society and affiliated groups held their annual meeting in Honolulu in early August, 1992. Before the meetings, I spent several days on the island of Hawai'i trying to do some field work, but extremely heavy rain made the work difficult. Surf on the southeast coast of Hawai'i was higher than I had ever seen it before. Local people seemed unconcerned, but several local residents said that it was one of the wettest summers they could remember. After the week-long meetings, and a few days of post-conference field trips, some two thousand or so visitors returned to their homes on the mainland. Three weeks later, on September 11, a disturbance that had originated south of the islands and had slowly grown to become Hurricane Iniki—the word *iniki* is defined (*P. & E.*) as "sharp and piercing, as wind or pangs of love." At full force, *Iniki* made landfall in the Poipu Beach area of southern Kaua'i. It is difficult to imagine that greater harm could have been done as this is one of the most heavily developed and frequently visited tourist areas of the island, with many private homes as well as luxury hotels lining the beaches. The estimated cost of damages due to *Iniki* probably approached 2.5 billion dollars! Also seriously affected was the National Tropical Botanical Garden, which lies a few kilometers to the west of Poipu Beach and only a scant few hundred meters from the coast. The storm continued northward across the island destroying large areas of forest as it went, leaving many scars that remain visible to this day, more than 11 years after the event.

Some days after the hurricane had passed, a group of botanists from the National Tropical Botanical Garden and the Bishop Museum began a survey of the damage caused by the storm. One of the places visited was the Blue Hole, which lies at the foot of the steep cliffs that make up the eastern flank of Mt. Wai'ale'ale. Among the debris that had accumulated was a tree in whose branches was entangled a plant that the botanists recognized as a species of *Lysimachia*, but not one that any of them knew. At that time, a young botanist named Ken Marr, who had had considerable experience in the Hawaiian flora, was working on his doctoral thesis on the Hawaiian species of *Lysimachia* in my laboratory at the University of British Columbia. Seeds from the newly discovered species were sent to Ken who successfully germinated them and grew them to maturity. As it turned out, the species was indeed new to science. Whether it was ever seen by early Hawaiians or how many of the plants still survive on the cliffs of Mt. Wai'ale'ale are not known. The new species has been named, appropriately we think, *Lysimachia iniki* Marr. Figure (6) shows the illustration that appeared in the description of this new species. For additional information on Hurricane Iniki, the interested reader may wish to log onto http://home1.gte.net/rhashiro/iniki.htm.

Nestegis sandwicensis—olopua, pua, ulupua

The little genus *Nestegis*, which belongs to the olive family, Oleaceae, consists of five species, three that are native to New Zealand, one that occurs in New Zealand and on Lord Howe Island, and one that occurs only on the Hawaiian Islands. When this

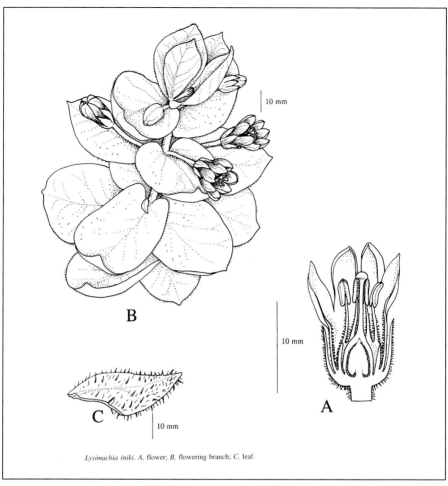

Lysimachia iniki. A, flower; *B,* flowering branch; *C,* leaf.

Fig. 6. Artist's drawing of *Lysimachia iniki* Marr reproduced from an article by K. L. Marr and B. A. Bohm (1997) in which the species was described for the first time.

species was first described, it was placed in the genus of the true olives, *Olea,* but was eventually rec-ognized as belonging to a separate genus. *Nestegis sandwicensis,* known as *olopua, ulupua,* or simply as *pua,* in Hawaiian, is illustrated in Plate (94). This specimen was found along the Awa'awapuhi Trail on Kaua'i. The wood of *olopua* is hard and durable and was favored by the Hawaiians for house construction.

Plate 94. *Nestegis sandwicensis,* a member of the olive family native to the Hawaiian Islands.

Plate 95. *Peperomia tetraphylla*, a widespread species in the Pacific Basin.

Plate 96. *Peperomia cookiana*, one of the native Hawaiian species.

Peperomia—'ala'ala wai nui

This next example takes us back into the realm of more familiar plants, at least for a sentence or two. The family Piperaceae, home to only eight genera, boasts nearly 3,000 species, among them *Piper nigrum*, the ground fruits of which the reader will know as black pepper. Two other species that are in common use, primarily in tropical countries and in the Pacific Basin, respectively, are *P. betel*, the betel nut, and *P. methysticum*, known as *kava*, *kawa*, or sometimes as *yangona*. Betel does not concern us in this book, but we will take a look at *kava* when we examine a few of the species brought to the islands by the early Polynesians.

Whereas *Piper* has no native species in the Hawaiian flora, its related genus *Peperomia*, 'ala'ala wai nui, is among the Hawaiian genera with the largest number of native species. Some authorities have estimated the number of species at about 50, but we will go with the *Manual* which lists 26 species, of which 23 are native, two are indigenous, and one is naturalized. The range of variation exhibited by the natives led to the suggestion by the authors of the *Manual* that there have been three, or possibly four, independent colonizations. The thousand species that constitute the genus are primarily pantropical in distribution. With this many potential donors, and the likelihood that seeds can be carried by birds, it is not difficult to imagine several introductions. I am unaware of any published studies examining this system, but it would seem to present an ideal opportunity to apply DNA sequencing techniques. Recall that the complexity of the Hawaiian lobelia relatives was originally thought to have been the result of several independent introductions but that DNA-based studies strongly suggest otherwise.

Plates (95) and (96) show two different species of *Peperomia* that are comparatively easy to find. The first, *P. tetraphylla*, was photographed along the boardwalk trail into the Alakai Swamp, Kaua'i. This species is pan-tropical in distribution and occurs widely in the Hawaiian Islands in diverse habitats. The second species, *P. cookiana*, was photographed along the trail at Kīpuka Puaulu on the Island of Hawai'i (going in the counter clockwise direction from the trailhead). Note the red stems of this species.

Pipturus—māmaki, māmake

Pipturus belongs to the family Urticaceae which most readers would recognize as being the home of the stinging nettle, members of the genus *Urtica*. The four species of *Pipturus* in the Hawaiian Islands are natives thought to have originated from a single introduction. The two examples here are very similar but occur on different islands, *P. alba* (Plate 97), which occurs on all islands was

Plate 97. *Pipturus alba,* a species widespread in the Hawaiian Islands.

encountered in the vicinity of Kīlauea Iki, on the island of Hawai'i, while *P. kauaiensis*, illustrated in Plate (98), occurs only on Kaua'i. This latter species is under cultivation in the Limahuli Garden. Fibers from the inner bark were utilized by Hawaiians for cordage and for preparation of a kind of coarse *kapa* (*tapa*) cloth (*Manual*).

Plate 98. *Pipturus kauaiensis,* a species native to Kaua'i.

Rosaceae—the rose family in Hawai'i

Roses, the genus *Rosa*, are likely one of the most familiar plants to most people. The family, Rosaceae, is also home to several other common plants including, blackberries, raspberries, strawberries, apples, and stone fruits such as apricots and peaches. Five genera of Rosaceae occur in the Hawaiian Islands, among them blackberries and strawberries. In this section we will meet all five genera and see how they serve as a microcosm of the Hawaiian botanical world.

Acaena, with about 100 species, is primarily a Southern Hemisphere genus but there is one species native to California; many others have been naturalized widely in the world. The genus is represented in the Hawaiian flora by one species, *A. exigua*, *liliwai* in Hawaiian. This species appears to be teetering on the edge of extinction. It was last collected in Kaua'i in 1869-1870, and on Pu'u Kukui on West Maui in 1957. Pu'u Kukui is not easily accessible and may still harbor populations of this species.

The second member of this group is also a plant familiar to most, the common strawberry, *Fragaria chiloensis*. As its name suggests, this species also occurs in Chilé, as well as in Argentina; it is also a common species of the Pacific Coast of North America, where it is strictly coastal. The Hawaiian Island plants have been accorded recognition as subspecies *sandwicensis*. Authors of the *Manual* note that the Hawaiian plants more closely resemble those that grow on the Juan Fernandez Islands, Chilé than those that grow in North America. This would seem to be an approachable research problem for someone interested in island travel! Returning to Hawai'i, we can expect to find this species of strawberry on East Maui and on Hawai'i. A comfortable place to see a population of *'ōhelo papa*, as the Hawaiians know it, is along the circular trail at the Kīpuka Puaulu on the Mauna Loa Road. Plants were seen a short distance from the trail sign along the right hand path. The Mauna Loa road leads to a lookout at 6,662' elevation where one gets an excellent view over the Kīlauea area. This is also the trailhead for the trek to the summit of Mauna Loa.

The next genus in our survey of Hawaiian Rosaceae is the indigenous *Osteomeles anthyllidifolia*, *'ūlei*, *u'ulei*, or *eluehe* (on Moloka'i), a mouthful in either language. This species also occurs on the Cook Islands and on Tonga with related species native to eastern Asia and Polynesia. I have seen fairly large mats of plants growing on lava along the Chain of Crater Road in Hawai'i Volcanoes National Park, but it grows in a variety of habitats on all of the main islands. Hawaiians called this plant *'ulei*, unless they were from Moloka'i where it was known as *eluehe*. The generic name is derived from the Greek *osteon* for bone, and *melon* for fruit, referring to the hard covering of the seeds. It has small, white flowers and characteristic small, shiny green leaves. Plate (99) illustrates the leaf arrangement. The wood of *'ūlei* is very hard and was used by the Hawaiians to make digging sticks and spears (Krauss,

Plate 99. *Osteomeles anthyllidifolia*, showing characteristic leaf array.

1993). They also made a two- or three-stringed musical instrument called an *'ukeke* which was strummed while holding one end in the mouth which served as a resonance chamber. Fruits of *'ūlei* were used to prepare a lavender dye.

Pyracantha angustifolia, which originated from southwestern China, is the fourth member of the list of island Rosaceae. Some members of this genus are cultivated for decorative purposes, which may explain why this particular one has become naturalized in Koke'e State Park on Kaua'i. The situation on Hawai'i described in the *Manual* is perhaps more disconcerting in that the plant is spreading near the Volcano town dump and in nearby abandoned fields. Given an opportunity of this sort, an aggressive weed can become a serious problem in a comparatively short period of time. This is especially true if its fruits are attractive to animals and thus gain wide dispersal.

The final example in this section is the genus *Rubus*, the blackberries. The *Manual* lists seven species, two of them native to the Hawaiian Islands, and five that are weedy and invasive species. The two native species are *R. hawaiensis*, which occurs on Kaua'i, Moloka'i, Maui, and Hawai'i, and the rare *R. macraei*, which occurs only on East Maui and on Hawai'i. The Hawaiians used the word *'akala* for both natives. The more common species, shown in Plate (100), was photographed on the eastern flank of Kīpuka Pu'u Huluhulu on Hawai'i. Authors of the *Manual* suggest that the resemblance of *R. hawaiensis* to the mainland *R. spectabilis*, the salmonberry, suggests that they have descended from a common ancestor.

Plate 100. *Rubus hawaiiensis*, the Hawaiian native blackberry.

The coffee family (Rubiaceae)

Although coffee is not a native species on the Hawaiian Islands, it nonetheless serves as a good introduction to this section because it is certainly the most well known member of its family, Rubiaceae. Another well known name in this family is *Cinchona*, the genus from which the antimalarial drug quinine was obtained. Rubiaceae are one of the largest flowering plant families consisting of 10,200 species in 630 genera (Mabberley, 1997), whereas the authors of the *Manual* make the family somewhat smaller, 6,500 species in 500 genera. Reflecting the size of the family is the substantial contribution Rubiaceae make to the Hawaiian flora, with a total of 66 species representing 16 genera. One genus, *Bobea*, with its four species, is native to the islands. The other native species represent genera otherwise well known in other parts of the world (with total species present in the Hawaiian flora in parentheses): *Coprosma* (13), *Gardenia* (3), *Hedyotis* (20), *Morinda* (1), and *Psychotria* (11).

     Coprosma consists of about 90 species the majority of which occur in New Zealand with other centers of diversity in Australia and New Guinea. Beyond these centers representatives of the genus can be found widely occurring in Borneo, Indonesia, Lord Howe Island, Norfolk Island, the Kermadec Islands, Pitcairn Island, and on the Juan Fernandez Islands of Chilé. An interesting feature of some members

of the genus, and the character that gives it its name, is the unpleasant aroma released when leaves or fruits are crushed. The name is based on the Greek "*kopros*" meaning dung, and "*osme*" meaning smell. Hawaiian species are much less evil smelling than species from other parts of its range, a feature thought to be related to the loss of defensive chemicals. One of the easiest species of *Coprosma* to find in the field is *C. ernodeoides*, which is characterized by grape-sized, shiny black berries (Plate 101). This species is known by several terms in Hawaiian: *'aiakanene, kukaenene, leponene,* and *punene,* all of which refer to the fact that the Hawaiian goose (*Nesochen sandvicensis*), the *nene,* eats the fruits. Plate (102) features a pair of these birds—fierce guardians of the park-

Plate 101. *Coprosma ernodeoides*, the *nene* bush.

ing lot at the end of the road on Kaua'i! The literal translation of *kukaenene* is goose dung, again suggesting an unpleasant odor, although I was unable to detect anything particularly dung-like, at least from crushed fruits. This plant can be found along the Halemau'u Trail on East Maui where a good display can be seen near the point where the latter trail begins its descent. On the Island of Hawai'i one can find this species growing on cinder fields in the vicinity of the Devastation Trail, or along the Mauna Loa Trail. Careful examination of several plants will reveal that two different flowers exist on this species; male and female flowers occur on separate plants—the species is said to be dioecious. This feature is illustrated in Figure (7).

Plate 102. A pair of *nene*, the Hawaiian goose, *Nesochen sandvicensis*, on Kaua'i.

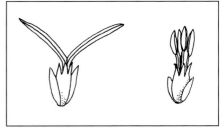

Fig. 7. Artist's diagram showing differences between female (left) and male (right) flowers as seen in *Coprosma*. The individual flowers are about a cm. long.

Coprosma elliptica (Plate 103) can be found growing along the trail beyond the Kalalau Lookout (Pihea Trail) on Kaua'i where it forms dense mats. This species resembles *C. ernodeoides* in its overall appearance except that its leaves are a paler green and its fruits, which are quite small, are yellow-orange. *Coprosma montana* can also be found along the Mauna Loa Trail among other places. This species is a small, stiff, upright tree that bears orange fruit. The leaves of these species appear to emerge from the stems in little tufts (Plate 104). This species also is dioecious, although occasional monoecious plants can be found.

Plate 103. *Coprosma elliptica*, another sprawling member of the genus.

Plate 104. *Coprosma montana*, a small tree here photographed near the Mauna Loa Trail head.

Fig. 8. *Hedyotis centranthoides* with detail of its fruit.

Coprosma offers an interesting challenge to the evolutionary botanist. It has been suggested by authorities that two, or possibly even three, independent colonizations may have occurred. *Coprosma ernodeoides*, with its black berries and certain other features, is quite distinct from the rest of the species on the Hawaiian Islands and is likely to have resulted from one of the colonizations. Opinions differ as to how the other dozen species are related, however, and whether they arose from a single introduction followed by diversification or whether there were two colonizations. Regardless of whether there were two or three events, it would be very interesting to see how the significant level of island endemism, which exists within this small group, came about. Other than *C. ernodeoides*, which we recognize as the odd man out, there are four species native to Hawai'i, three that occur only on Kaua'i, one known only from Moloka'i, and another only from O'ahu. The remaining species occur on several islands but, with one exception, not on Kaua'i. In addition to determining what the relationships are among the island species, it would be necessary to establish where the colonists might have come from.

As mentioned above, the genus is widely dispersed in the Pacific Basin ranging from southeastern Asia to the Juan Fernandez Islands, Chilé.

Returning to our survey of Rubiaceae we can look briefly at the genus *Hedyotis*. Twenty native species are listed in the *Manual*, several of which are endangered, rare, or extinct. The only species of this genus for which I have an illustration is *H. centranthoides*, which is characterized by widely spaced, thick, purple-green leaves, a drawing of which appears as Figure (8). The tiny, pale yellow-green flowers of this species are illustrated in Plate (105). This plant is common along the power line access road that parallels Route 200 (the Saddle Road) on Hawai'i, access to which can be found at about the 19 mile marker. This is also a good area in which to see other native species including some ferns, a species of *Dubautia*, and the lichen *Stereocaulon vulcani* which occurs widely distributed over lava. This organism, which is white or grayish (Plate 106), serves an extremely useful function because of its capacity to reflect much of the sun's energy, thus preventing the temperature of the underlying black lava rocks from rising to a level lethal to seeds or fern spores.

Plate 105. *Hedyotis centranthoides* flowers are tiny and yellow-green but turn black on drying.

Plate 106. *Stereocaulon vulcani,* the lava lichen.

Plate 107. *Hedyotis terminalis,* an extremely variable species with tiny, white flowers.

A species of *Hedyotis* that prefers wetter, often swampy areas, is *H. terminalis*, known as *manono* in Hawaiian. The plant photographed for Plate (107) was found in a swampy field beside the road to Hawai'i Volcanoes National Park on Hawai'i at about 2,500' elevation. This plant is not included as one of the island's spectacular botanical beauties with its small, unimpressive flowers and black fruits! It is botanically interesting because of its extreme variability: the authors of the *Manual* describe *manono* as being perhaps the most variable species on the islands other than possibly *Metrosideros polymorpha*. To illustrate the level of variation seen in specimens now considered to fall within the limits of the species, there is an entire page (small print) in the *Manual* dedicated to listing other names by which speci-

Plate 108. *Morinda citrifolia,* the *noni* or Indian mulberry was brought to the islands by the Polynesians.

Plate 109. *Coffea arabica,* commercial coffee plant.

mens have been listed. It is not surprising that the authors suggest that this might be a challenging subject for detailed investigation.

Morinda, a genus of the Old World tropics, is represented in the Hawaiian Islands by two species, the rare, native M. *trimera,* and the Polynesian introduction M. *citrifolia,* known locally as *noni* (it is known in Asia as the Indian mulberry). This species is illustrated in Plate (108). The most important uses of this plant seem to be the production of dyes for dying tapa, red from the bark and yellow from the roots. Juice of the fruit was used in the preparation of a medicine for tuberculosis.

Two species of *Coffea* are cultivated in the islands, C. *arabica* and C. *liberica,* both of which are native to Africa. Coffee was introduced to the Hawaiian Islands early in the 19th century, one authority says 1813, another 1823. Coffee plantations on the leeward side of Hawai'i are home to "Kona" coffee. There is also extensive cultivation on the southern coast of Kaua'i, which can be seen when driving west along Route 50. The characteristic berries of this species are attached directly to the stem (Plate 109).

Rutaceae

The most familiar member of Rutaceae is the genus *Citrus*, which the reader knows of course as lemons, limes, oranges, grapefruits, and numerous hybrids. Four genera of Rutaceae occur in the Hawaiian Islands, one of which, *Platydesma*, with four species, is native. The other three genera are either naturalized (*Flindersia* with one species) or have native species on the islands but are well represented elsewhere (*Pelea* and *Zanthoxylum*). We will focus our attention on *Pelea* in this section. Up to a few years ago, *Pelea* was considered to consist of 47 species native to the Hawaiian Islands with an additional two native to the Marquesas. Recent detailed studies of *Pelea* and related genera revealed a significant level of similarity between *Pelea* and *Melicope*, a genus that ranges from the Indomalaysian peninsula through Australia to New Zealand. Unfortunately, because of botanical naming rules, the newer name, *Pelea* as the genus name had to go. How this nomenclatural activity will be received by *Pele*, the Goddess of Volcanoes, remains to be seen. The name change does not affect the native status of the Hawaiian species and it establishes the subgenus *Pelea*—within *Melicope*—making it a native subgenus.

The species that we will look at here, *Melicope* (*Pelea*) *anisata*, has a number of features of interest, one of which can be seen in Plate (110). The four-parted fruits of this shrubby vine, whose common name is *mokihana*, is much prized for the making of *lei*. The term "anisata" refers to the strong anise-like aroma of all parts of the plant. The intensity of the aroma varies from organ to organ and from tree to tree, but it is still a very useful feature in helping to identify the species, although others do have an anise-like aroma as demonstrated by chemists at the University of Hawai'i (Scheuer 1955, Scheuer and Hudgin, 1964), who identified the anise-like odor component in M. (*P.*) *christophersenii* as well. There is a serious drawback to the use of *mokihana lei*, however; these plants contain chemicals that, when exposed to sunlight, can result in contact dermatitis. I have seen photographs of people with a ring of welts around their neck matching the places where the fruits came in contact with their skin. A group of workers in the Botany Department at the University of British Columbia (Marchant et al., 1985) determined that the compounds responsible for this effect are well known coumarin derivatives that become activated by sunlight and bind with certain skin components forming complexes against which the body reacts.

Plate 110. *Melicope* (*Pelea*) *anisata*, fruits of *mokihana*.

Distribution of Rutaceae in the Hawaiian Islands

It is interesting to look at the distribution of the native Hawaiian species in a little more detail. We have seen other instances where species are native to individual islands, and in fact to particular regions on certain islands, such as with the two species of *Silene* restricted to young lava described earlier in this chapter. A more detailed examination of the distribution of *Melicope* (*Pelea*) reveals a striking level of endemism within the archipelago with 36 species restricted to single islands. Of the 10 species that occur on two or more islands only a few could be described as common. (This only adds up to 46 species; the 47th has been discounted in this summary owing to uncertainty as to which island it came from, and it is extinct.) An added problem in assessing the group is the troublesome fact that five of the single island natives (three from Kaua'i and one each from Maui and Moloka'i) are known only from the type collections. If we take these to be legitimate species, it is interesting to note that the number of native species decreases with decreasing age of the respective islands, thus, Kaua'i has 14 native species, O'ahu 10, Maui Nui 9 (Maui 6, Moloka'i 2, Lāna'i 1), and Hawai'i 3. This apparent correlation with age of island is generally taken to suggest that the greatest diversification occurred on the oldest island followed by colonization of younger islands. The problem of sequential colonization of islands is a topic frequently discussed in cases where species occur on more than one island. Only with detailed information on the evolutionary relationships among species on neighboring islands, however, is it possible to determine the direction of inter-island colonization with any degree of certainty. Normally the direction is from older to younger, but colonization of older islands from younger islands is known to have occurred in a few instances.

Sandalwood—legitimate and bastard

The history of the Hawaiian Islands is intimately connected with the development of agriculture, particularly pineapple and sugarcane. Although the vagaries of commerce have shifted emphasis on these familiar products, extensive acreage of both continue to be maintained. In recent years coffee has also become a major crop. Although not as noteworthy as any of the three big crops just mentioned, sandalwood played an interesting role in Hawaiian commerce in the early 19th century. Whereas extensive tracts of land were cleared of native vegetation to make way for pineapple and sugarcane fields, it is the native vegetation itself, sandalwood forests, that was harvested for profit in the islands, with devastating results.

Santalum (Santalaceae) is a genus of about two dozen species that range from India through southeastern Asia to Australia with representatives in the Hawaiian Islands and one each on the Juan Fernandez Islands, Chilé, and in the

Marquesas. Sandalwood is thought to be extinct in Chilé owing to heavy over-harvesting. *Santalum album* has been cultivated in India for many years as a source of fragrant wood used for the manufacture of chests, other items of furniture, incense, and a perfume. It was also the wood used for Buddhist funerals. Aromatic heartwood is a feature shared by other members of the genus, including, *S. freycinetianum*, one of the Hawaiian species, called *'iliahi* by the Hawaiians. Discovery of sandalwood in the Hawaiian Islands led to establishment of trade in the wood in 1791 by Captain John Kendrick, a fur trader from Boston. In the early 1800s another group of American traders, whose normal business involved shipping furs from the west coast of North America to China, arranged a partnership with King Kamehameha I (recall, he had been present at the death of Captain Cook) whereby he would get 25% of the profits from the sandalwood trade. Their plans were interrupted by the War of 1812 and eventually the deal fell apart without anyone making a great deal of money. An association between American and Russian business interests emerged, however, that exploited the ready availability of sandalwood and the harvest began in earnest. A detailed account of the fascinating political interactions underlying this arrangement, including the intriguing possibility that Kaua'i might have become Russian territory, can be found in Gavin Daws' history of the islands. A more personal view of the sandalwood trade, and life in the islands in general, comes form the journal of Charles H. Hammatt, a young man from Boston involved in the sandalwood trade who lived in the islands during the period 1823 to 1825 (Wagner-Wright, 1999).

The devastation of sandalwood forests might be better appreciated with a few numbers. In the period 1821-1822, 1.8 million kilograms (ca. 4 million pounds) of sandalwood were shipped to China; in 1822-1823, 1.1 million kilograms. It has been estimated that a shipload of logs required about 6,000 trees to be harvested. The sandalwood forests were not the only casualty of these harvests. The heavy work of cutting the trees and transporting the logs fell to the commoners, many of whom were severely injured or killed in the course of their toils (Kepler, 1998). Dr. Kepler also describes Moloka'i's "sandalwood hull," a hollow in the shape of the ship's hull in which the cut logs were to be transported and thus measured for shipment.

By 1831, as noted in the journals of Dr. F. J. F. Meyen, a young Prussian naturalist, the trade in sandalwood was already significantly reduced, and by the time the United States Exploring Expedition arrived on the islands in 1838, only small shrubs were seen. By the mid 1900s sandalwood forests on O'ahu had been decimated. Fortunately for the species, small plants were safe from exploitation because the aromatic constituents only accumulate to commercially useful levels in older trees. Also, a few small populations in inaccessible sites were spared. The reader might find it interesting to learn that one method used for finding sandalwood trees was to set

Plate 111. *Santalum freycinetianum*, the Hawaiian sandalwood.

Plate 112. *Exocarpus menziesii*, an odorless member of the sandalwood family.

fires and smell the smoke for the characteristic sweet aroma. Presumably a "positive" test would lead to extinguishing the fire so that the remaining trees could be cut. One wonders if the harvesters left a forest to burn if it didn't smell right. The contribution of wholesale burning to destruction of extensive lowland forests was discussed in *Alteration of Native Hawaiian Vegetation* by the Hawaiian botanists Linda Cuddihy and Charles Stone (1990).

Returning to Hawaiian *Santalum*, we see from the *Manual* that four native species are known, all of which are rare. The level of morphological variation within each of these species is such that four names may not adequately reflect all the differences that seem to exist. Plate (111) shows the characteristic flower presentation of *S. freycinetianum*. This small tree was found in wet forest along the Awaʻawapuhi Trail at Kokeʻe. Although heavily victimized, as described above, this species occurs on all of the main islands but Hawaiʻi. *Santalum haleakalae* is native to sub-alpine scrub on East Maui. A few trees of this species can be found beside the

Plate 113. *Myoporum sandwicense,* false or bastard sandalwood.

road near the Visitor's Center in Haleakalā National Park. *Santalum paniculatum* occurs in a variety of settings ranging from dry lava fields to wet forests on Hawaiʻi.

A second genus in Santalaceae also occurs on the Hawaiian Islands. *Exocarpos*, represented on the islands by three native species, is a genus of about two dozen species that occur in southeastern Asia, Australia, and in the Pacific Basin. *Exocarpos menziesii*, shown in Plate (112) was found at about 7,200' elevation along the Mauna Loa Trail. This is a striking plant with dark purple stem tips and a broom-like appearance. The specimen pictured was characteristically nearly leafless. From a distance, it is possible to mistake the purple coloration as the brown twigs of some dead shrub, but closer inspection leaves no doubt that it is alive. Owing to its scattered occurrence and its unusual growth form, it is likely that *Exocarpos* could be one of the first-time visitor's most exciting finds on this trail.

There also exists on the Hawaiian Islands, and other islands in the Pacific Basin, and on Mauritius, New Guinea, New Zealand, Australia, and eastern Asia, members of the genus *Myoporum* (Myoporaceae). The sole representative of *Myoporum* on the Hawaiian Islands is M. *sandwicense*, known locally as *naio*. This species came to be called "bastard sandalwood" because it too has a heartwood with a pleasant aroma. After the supply of true sandalwood was exhausted, efforts were mounted to sell M. *sandwicense* to the Chinese. Shipments were rejected however. Young *naio* trees, among other species, were used by early Hawaiians as house posts, and shuttles used for making netting were fashioned from thin *naio* branches (Krauss, 1993). Plate (113) features flowers of *naio* found on Kīpuka Puʻu Huluhulu.

Scaevola—naupaka

At least one member of the genus *Scaevola*, *naupaka* in the Hawaiian language, would be among the easiest of the Hawaiian indigenous species to find: *S. sericea* (Plate 114) — called *S. taccada* by Howarth et al.—is a strand plant frequently used for beach stabilization, where it doubles as an attractive species in its own right. Characteristic of these species

Plate 114. *Scaevola* (*taccada*) *sericea*, *naupaka*, a strand plant widespread in the Pacific basin.

of *naupaka*, with the exception of *S. glabra*, is what some have likened to a "half flower," a feature clearly visible in the plate. Hawaiian legend has it that the full flower was torn in half during a lovers' quarrel and that the situation will not be rectified until the full structure is restored (see Kepler (1998) for the full story). As a taxonomic aside, it might be pointed out that returning this plant's flowers to the "full" status might well require its removal from the plant family (Goodeniaceae) in which it has dwelt since its discovery! *Scaevola*, with about 100 species, is represented in the Hawaiian Islands by eight natives and one that occurs widely in the Pacific Basin (*S. sericea*). The genus is mainly Australian, where as many as 70 native species occur, but is also represented in New Caledonia and in Irian Jaya. Other islands in the Pacific Ocean can also boast their own native members of this genus: the Marquesas have two, and the Society Islands, Samoa, and Fiji each have one.

The rest of the Hawaiian species, save one, have open flowers that resemble the beach plant. *Scaevola glabra*, illustrated in Plate (115), is easily distinguished from other Hawaiian species because of its tubular yellow flowers, obviously an adaptation to bird pollination. It appears to be most closely related to a tubular flowered species, *S. coccinea*, from New Caledonia. *Scaevola glabra* can be found on Oʻahu and Kauaʻi where it occurs in wet forests. The photograph was taken on the board walk to the Alakai Swamp on Kauaʻi.

The relationships among the Hawaiian species has attracted a good deal of attention over the years. In addition to the obvious difference in flower shape, these species also differ in chromosome number, which speaks loudly about relationships. In addition to tubular yellow flowers, *S. glabra* is a tetraploid species, which means that it has a double set of chromosomes, in this case $2n = 32$. The beach *naupaka S. sericea* (*S. taccada*) is also tetraploid but the morphological features of these two

Plate 115. *Scaevola glabra,* one of the endemic species on the Hawaiian Islands.

species are so different that a close relationship isn't likely. The remaining species on the Hawaiian Islands, however, have the normal diploid number of chromosomes, $2n = 16$. The differences in floral structure and chromosome numbers, as well as other features, have led to the suggestion that three independent colonizations must have occurred to account for the presence of *Scaevola* on these islands. A recent study by Dianella Howarth, currently at Yale University, and colleagues (2003) using DNA sequence data clearly indicates that this is in fact the case. The data pointed to three relationships for the present island species: (1) *S. sericea* (*S. taccada*) to a group of species from Polynesia; (2) *S. glabra* to a group in Australia; and (3) the rest of the species possibly to diploids from the Americas.

It is interesting to look at the beach *naupaka* in a little more detail. Its presence on the beaches of most Pacific islands indicates that its seeds can be readily transported by ocean currents, which is known to be the case. Distinct from inland species, seeds of beach *naupaka* include a corky layer that provides excellent buoyancy. It has been suggested by L. Van der Pijl (1972), an authority on dispersal mechanisms in higher plants, that seeds of species such as the beach *naupaka* germinate on a beach after a period of time of flotation in sea water followed by a bath of fresh water (rain). Seeds of the other *naupakas* on the Hawaiian Islands are apparently not capable of surviving long time immersion in sea water and are thought to have arrived on the islands by other means, most likely birds. These special circumstances may explain why *S. sericea* is restricted solely to life on the beach and all other *naupaka* occur elsewhere. Why beach *naupaka* has not undergone speciation, as the other diploid species appear to have done, is likewise unknown.

Before moving on it is useful to emphasize the significance of these studies on the *naupaka*. So far as is known at the moment, *Scaevola* is the only genus of native Hawaiian plants that has **not** resulted from a single colonizer. DNA-based research clearly point to the three different forms of *Scaevola* having resulted from independent colonization events. The picture may be even more complex when one considers the discovery of a fourth type of *Scaevola* on the islands. Warren Wagner—the *Manual* Wagner—(1996) described a new species of *naupaka,* *S. hobdyi,* from a single specimen that had been collected on West Maui. The new species is seen as sufficiently different from all of the other types on the islands, that a fourth colonization event may have been involved. Baring the discovery of additional members of this species on the islands, or possible relatives elsewhere, the role that this species has to play in the *naupaka* story must remain a mystery.

Sicyos

Sicyos belongs to Cucurbitaceae, the family that includes such familiar plants as cucumbers, gourds, melons, squash, and zucchini. *Sicyos*, the Greek word for cucumber, is a genus of 50 species with a range that includes the Americas, islands in the southwestern Pacific Ocean, New Zealand, Australia, and the Hawaiian Islands. The genus in Hawaii consists of 14 native species, one of which is thought to be extinct and two that are listed as rare. The species illustrated below, *S. alba*, one of the rare species, was found growing on the southern slope of Pu'u Huluhulu on Hawai'i. *Sicyos alba* can be readily

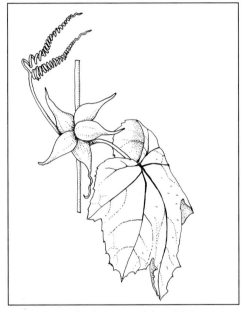

Fig. 9. *Sicyos alba* showing tendrils and fruit structure as the plant might be seen climbing a tree.

Plate 116. *Sicyos alba* leaves.

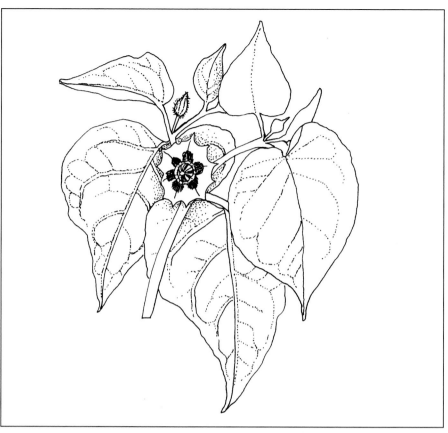

Fig. 10. *Physalis peruviana*, the so-called Cape gooseberry. The common name is wrong in two respects; the plant is native to Peru, not the African Cape, and it is not a gooseberry. Gooseberries are related to the currant, whereas *Physalis* belongs to Solanaceae, the potato family.

recognized by its fruit which, when ripe, consists of a group of white, comet-shaped structures attached at their bases as illustrated in Figure (9), and by its long, tightly coiled tendrils by which it attaches itself to other plants. The characteristically lobed leaves in mottled sunlight are illustrated in Plate (116).

Puʻu Huluhulu is a *kīpuka*, an island of forest surrounded by lava flows. This particular *kīpuka* is a post shield stage vent of Mauna Kea surrounded by lavas from Mauna Loa. Kīpuka Puʻu Huluhulu can be found at about 6,500' elevation at the junction of the Saddle Road (Rt. 200) and the Mauna Kea road and near the turnoff to the Mauna Loa Observatory Road. The *kīpuka* serves as a small nature preserve that is supposedly protected from feral animals. (It didn't seem secure on my last visit!) There is a hunter check-in station at the parking lot from which brochures are sometimes available (Anonymous).

Also in flower on the *kīpuka* (middle May) were bastard sandalwood, also known as *naio* (*Myoporum sandwicense*) (Plate 113), the Hawaiian blackberry (*Rubus* hawaiiensis) (Plate 100), a weedy species of *Bidens*, and the introduced Cape gooseberry (*Physalis peruviana*), locally known as *poha*, (Figure 10). An excellent example of an invasive weed is the healthy growth of bracken fern (*Pteridium aquilinum*) on the northern slope of the *kīpuka*; the invasion was likely to have occurred following a fire in the 1950s that destroyed much of the native vegetation on that slope (Plate 117).

Smilax—cat- or green-brier

Some years ago when I was doing field work in the New England states it was necessary to be aware of *Smilax* in order to avoid entanglement with its unforgiving recurved hooks. As is the case with other Hawaiian native

Plate 117. Bracken fern (*Pteridium aquilinum*) on the slopes of Pu'u Huluhulu.

species, the blackberries for example, armaments such as barbs, hooks, or spines are often reduced or lost altogether compared with their mainland relatives. Another example is *Smilax melastomafolia, hoi kuahiwi* in Hawaiian, a native species that occurs in wet forests and bog margins on all of the main islands. This species is a climber (liana), as are many of its mainland relatives, but has significantly reduced spines. This plant was photographed along the trail to the Alakai Swamp (Plate 118). It is easy to recognize with its shiny leaves—they appear almost to have been waxed—and its climbing growth habit. The specific epithet refers to the similarity of leaf venation to the pattern that is characteristic of members of Melastomataceae, a

Plate 118. *Smilax melastomatifolia.*

completely unrelated family some of whose members we will meet in Chapter Three (e.g., *Tibouchina*). *Smilax* consists of about 300 species worldwide with one native to Fiji, *S. vitiensis*, that shares certain features with the Hawaiian species.

This next set of examples are species of flowering plants that occur naturally in the Hawaiian Islands as well as on other islands. A few examples of Polynesian introductions are included here as well.

Dianella

Liliaceae, the lily family, is represented on the Hawaiian Islands by three native species of *Astelia*, one indigenous species of *Dianella*, and one naturalized species of *Hippeastrum*. The indigenous *Dianella sandwicense*, the 'uki'uki or simply 'uki, is fairly widespread in a variety of habitats on all of the main islands except Ni'ihau and Kaho'olawe. It also occurs in the Marquesas. Plate (119) shows this species with its attractive blue fruits.

It is of no pressing importance to us in the present context, but the reader might be interested to learn that a re-examination of Liliaceae a few years ago by a group of botanists resulted in all three of the Hawaiian members being relocated to other families: *Astelia* to its own family Asteliaceae, *Dianella* to Phormiaceae, and *Hippeastrum* to Amaryllidaceae. About the only impact this rearrangement would have on the Hawaiian flora is a change in the census of families represented in the islands. These sorts of rearrangements may reflect more up-to-date inter-family evolutionary relationships, but for the amateur botanist they provide little more than an annoyance by having tinkered with more or less well known names. The new relationships do offer targets for research, however.

Plate 119. *Dianella sandwicense,* a lily also known in the Marquesas.

Dodonaea

Dodonaea viscosa, called *aʻaliʻi* in Hawaiian, is a member of the family Sapindaceae. *Dodonaea* consists of about 65 species, the majority of which are native to Australia.

The Hawaiian species is pantropical in distribution and occurs on all of the main islands except Kahoʻolawe. This species is abundant and can be found, among other places in alpine scrubland, along the Halemauʻu Trail on Haleakalā, along the Mauna Loa Trail, and in the vicinity of the Volcano Observatory. Male and female flowers, which are small and inconspicuous, occur on different plants. The two- to four-angled fruits, however, are quite attractive and appear in various shades ranging from red to reddish-brown and almost purple-brown in some places. The fruits, which are classified as capsules, were used to prepare a red dye. Fruits and leaves are both used to make *lei*. Plate (120) shows the fruits of this species.

Plate 120. *Dodonaea viscosa,* featuring its very attractive fruits.

Styphelia

Styphelia tameiameiae is the only member of Epacridaceae, primarily a Southern Hemisphere family, in the Hawaiian Islands. (In the view of some botanists this and a dozen or so other species are sufficiently different to justify recognition as the separate genus *Cyathodes*.) Known as *pūkiawe*, *S. tameiameiae* is one of Hawaiʻi's abundant indigenous species. It is easily found along the Halemauʻu Trail among many other places throughout Haleakalā National Park. It is also abundant along the Mauna Loa Trail. The flowers are tiny, but the globular fruits, which range from white through pink to red, are easy to see (Plate 121). Leaves are small and are frequently used, along with fruit, in making *lei*. Smoke from burning *pūkiawe* was used to smudge high-ranking chiefs who wished to mingle with the common folk. The specific epithet *"tameiameiae"* is a Latinized form of Kamehameha, thus honoring the name of a line of Hawaiian kings.

Plate 121. *Styphelia tameiameiae.*

Freycinetia arborea

Freycinetia consists of about 180 species and is represented on the Hawaiian Islands by *F. arborea*, 'ie'ie or 'ie. This species occurs otherwise in the Marquesas, the Society Islands, the Austal Islands, the Cook Islands, and in New Caledonia. The specific epithet "arborea" suggests that this species is a tree, which it is not. It is capable of climbing trees, however, and can form an attractive display that may suggest that it, rather than its host, is the center of attraction. It

Plate 122. *Freycinetia arborea* female "cones."

can also be found as a creeper when no trees are available. Plate (122) shows a specimen of *F. arborea* that was encountered along the Boy Scout Trail on West Maui. *Freycinetia* is a member of Pandanaceae, the screw pine family. Louis de Freycinet, after whom the genus was named, was captain of the *Uranie*, and commander of the French exploring expedition of 1817-1819. On board was the botanist Charles Gaudichaud-Beaupré who collected on Hawai'i, Maui, and O'ahu in 1819. The full name of this plant, *Freycinetia arborea* Gaud., therefore recognizes contributions by both de Freycinet (the genus name) and Gaudichaud (the authority name).

Pandanus tectorius

Pandanaceae consists of three genera: *Freycinetia*, which we just saw, the tropical *Sararanga* with two species, and *Pandanus*, which we can look at briefly now. *Pandanus* consists of as many as 700 species (Mabberley, 1997) and enjoys a wide tropical distribution. The only species in the Hawaiian Islands is *P. tectorius*, the so-called screw pine. It gets that name because of its stilt roots that appear to emerge from the

Plate 123. *Pandanus tectorius,* the screw pine.

stem in a more or less spiral fashion. The Hawaiians refer to the plant as *hala* or *pū hala*. Plate (123) shows the typical long, strap-like leaves and the large fruit. It is not uncommon for gullible tourists to be told that this is actually the "pineapple tree." The leaves, *lau hala*, were used to make floor mats and other plaited objects such as hats, purses, or bags (Krauss, 1993). Root fibers have been used as cordage and the seeds can be eaten.

Morning glories

The beach morning glory, *pōhuehue*, *Ipomoea pes-caprae* subsp. *brasiliensis*, is a prominent member of the strand community throughout the Pacific. This plant can be found above high water line on most undisturbed beaches throughout the Hawaiian Islands including islands to the northwest of the main group. The photograph for Plate (124) was taken on beach dunes at Polihale State Park, illustrated in Plate (18).

Plate 124. *Ipomoea pes-caprae,* the beach morning glory, widespread in Pacific islands.

This indigenous species belongs to the family Convolvulaceae, members of which include the morning glories and are widely cultivated for floral displays, while others are valued foods. The most common member of this genus used for food is *I. batatus*, known as the sweet potato. Sweet potatoes were among the plants brought to the islands by the early Polynesians who used them as a dietary source of starch.

The early Hawaiians twisted the long stems of *pōhuehue* into a rope that was used in making fish nets and fish traps. Medical uses included application of a preparation of mashed vine to a sprained joint, and administering a preparation from mashed roots as an effective cathartic (Krauss, 1993).

The *Manual* lists 14 species of *Ipomoea* on the islands, only one of which is native. *Ipomoea tuboides*, the Hawaiian moon flower, grows on *a'a* lava or rocky arid sites on all of the main islands. I have not seen this species in the field but according to the *Manual* it can be recognized by its long, white, tubular flowers.

Vitex rotundifolia

Vitex rotundifolia, a member of Verbenaceae, is another very attractive beach plant that occurs widely, ranging from Mauritius and Sri Lanka in the west, through eastern Asia and Australia, to many of the Pacific islands. It occurs on all of the main Hawaiian Islands except Kaho'olawe. It is well-rooted in sand dunes and is often planted to take advantage of its capacity to bind sand. Plate (125) illustrates this species in its native habitat at Polihale State Park, where it grows in the vicinity of *Ipomoea pes-caprae*. The flowers are pale blue and the leaves pleasantly aromatic. Among other names, the Hawaiians called this plant *kolokolo kahakai* or *hinahina kolo*. The word *kolo* means to creep or crawl in reference to this plant's creeping growth habit. The term *hinahina* is used to describe a number of species that are characterized by grayish or white tomentum on leaves, as in *Argyroxiphium* (silversword) and *Artemisia*. This plant finds favor in native landscaping on the islands.

Plate 125. *Vitex rotundifolia,* widespread beach vitex.

Waltheria indica

I met *Waltheria indica* in a somewhat roundabout way. While I was taking pictures of the two plants above, my wife was showing some of her drawings to some folks (from Līhu'e) who were camping on the beach. We spent some time talking to them about uses of native species—including a plant that was growing near their tent that one of the gentlemen said was used for preparing a "tonic." He called it *'uhaloa* although noting that it was called different things by different people and on other islands. Several days later, as we were preparing to fly from the Honolulu airport, I was asked to open my camera bag for inspection. The young woman in charge of this security check

Plate 126. *Waltheria indica,* a medicinally useful species possibly introduced by Polynesians.

asked why I was carrying a copy of *Hawaiian Herbal Medicine* (by June Gutmanis) in my bag. After telling her about our work for this book and our interest in uses of local plants, she informed us that her father, from Samoa as I recall, used a tea made from leaves of a plant called *'uhaloa* as a cure for sore throat and that it was a very well known medicinal plant. Working backwards from the Hawaiian name, I found that we had stumbled upon (quite lit-

erally) *Waltheria indica*, a member of the family Sterculiaceae. Plate (126) features this species with its typical small yellow flowers. In addition to life as a low, sprawling beach plant, this species can also attain the status of small shrub. The *Manual* tells us that a bitter preparation of the inner bark and roots was used as a sore throat remedy. The *Manual* lists this species as possibly indigenous. It was first reported from the Hawaiian Islands in 1779, suggesting that if anyone brought this pantropical plant to the islands, it would likely have been the Polynesians. That would not be at all surprising since its medicinal properties appear to be widely known.

Kukui

Seeds of *Aleurites moluccana* were brought to the Hawaiian Islands by the Polynesian colonizers who were well aware of the many uses to which parts of this tree could be put, as well as its connection with the hog god Kamapua'a, for which it was a form (Handy et al., 1991). The species, known as *kukui* (or *kuikui* at one time) or candlenut, is a member of the Euphorbiaceae, a large (ca. 7,500 species) cosmopolitan

Plate 127. *Aleurites moluccana,* leaves and flowers of the *kukui* nut.

family well represented in tropical and subtropical areas. In addition to trees that escaped from cultivation, and others that were purposefully planted, this species now occupies mesic valleys on all of the main islands except Kahoʻolawe. The fruits of the *kukui* are the size of chestnuts, very hard, and oil rich. The nuts are attractive and are used to make permanent *lei.* Leaves and flowers are illustrated in Plate (127). *Kukui* trees are very easy to identify at a distance because of their pale gray-green leaves which can be seen reflecting sunlight in the photograph in Plate (128).

The common name candlenut refers to the use of the nuts as a source of light provided by the slow-burning oil. The oil was also extracted on a commercial scale

Plate 128. *Aleurites moluccana,* trees with *ti* (*Cordyline*) in the foreground.

many years ago for use as a drying oil for the manufacture of paints and varnishes as is the case with tung oil, a product obtained from fruits of the related *Aleurites cordata*. The oil also has strong laxative properties from which use in Malaysia has come the highly descriptive name "1,2,3-nut," in reference to the speed at which the *kukui* nut oil can bring about the desired (?) effect. (This common name appears not to be used in the Hawaiian Islands.) With regard to its purgative action, the oil from *kukui* nuts is similar to castor oil obtained from *Ricinus communis*, also a member of Euphorbiaceae. Incidentally, *Ricinus* is a widespread weed that had become naturalized in the Hawaiian Islands before 1819. When roasted and mashed with salt, *kukui* nuts provided a favorite relish to which peppers were sometimes added for additional zest.

In addition to the uses of *kukui* just noted, its oil was used to polish and waterproof food dishes and as a house light, either by burning the nut itself or by using a *tapa* wick in a stone bowl of oil. A torch consisting of nut shells was used at night to attract reef fish. Nuts were also chewed by fisherman and the mass spit onto the surface of the water which smoothed the ripples and made octopus burrows on the bottom easier to see. *Kukui* wood was occasionally used in the construction of gunwales for canoes when the more favored wood, from species of *Bobia* (ʻahakea), was not available. Several parts of the *kukui* tree provided dyes, gray from fruit husks, charcoal from burned nuts, and a reddish pigment from inner bark that was used to render fishing lines less visible. The soot from a burned husk was used to paint a black line on the tongue as a part of the mourning process. Many other uses of this extremely versatile tree are discussed in detail by Abbott (1992), Krauss (1993), and Kepler (1998). It shows no disrespect to suggest that *kukui* might be the botanical equivalent of the pig whose every part was used except the squeal.

In the foreground of Plate (128) are several *ti* plants. *Ti*, or *kī* is *Cordyline fruticosa*, a member of the agave family (Agavaceae). *Cordyline* was brought to the islands by the Polynesian colonists who used leaves for food wrappers (used in *luau* preparations), house thatches, and items of clothing. Roots provided a food stuff and an alcoholic beverage. Children used large leaves as sleds for mud slides.

Piper methysticum—Kava

Piperaceae consists of eight genera, one of which, although quite large, contains one of the most familiar plants known to mankind, *Piper nigrum*, whose ground fruits provides the black pepper of commerce. A second species, of particular interest to us in this chapter, is *P. methysticum*, whose common names include *kawa* and *yangona*, and in the Hawaiian Islands, ʻawa, *pūʻawa*, and *kava*. *Piper methysticum* is a sterile cultigen (cultivated variety) derived from *P. wichmannii*, a sexually reproducing species native to New Guinea and Vanuatu. *Kava* has been a companion of people of the Pacific Islands for a very long time and is certain to have accompanied them on their voyages of discovery. The roots and lower bits of stem, either fresh or dried, are used to prepare a drink, often used ceremonially, but increasingly as a social beverage. The effect of the drink is a mild euphoria (technically described as a narcotic sedative). *Kava* is cultivated on nearly all Pacific Islands including the Hawaiian Islands and often can be found in most botanical gardens and in private gardens as

Plate 129. *Piper methystichum*, kava, much cultivated in the Pacific basin.

well. Detailed studies of variation in chemical composition, preparation and other features have been described by V. Lebot and J. Lévesque (1989). Plants under cultivation can be seen in the Limahuli Garden. Plate (129) illustrates the *kava* plant with its characteristic jointed stem.

Saccharum—Sugar cane

Cultivated forms of sugar cane were originally brought to the Hawaiian Islands by the Polynesian colonists. Sugar cane was an extremely important plant to early Hawaiians as evidenced by their knowledge of some 40 varieties. Perhaps the best known species of *Saccharum* is *S. officinale*, which is a complex assortment of hybrids originally domesticated in New Guinea (Mabberley, 1997). Botanically "pure" sugar cane is likely lost in the history of agriculture, although reconstruction of a sugar genealogy might be possible using the modern tools of molecular genetic analysis. Sugar, of course, has played a major role in the development of the agricultural industry in the Hawaiian Islands. The complex politics of sugar can be appreciated by reading Gavin Daw's (1974) writings on the subject.

Sugar plantations are very common in the Hawaiian Islands where it is often possible to see a field in some stage of the process, from preparation through to the burning of the field preparatory to the harvest itself. Great trucks laden with blackened sugar cane stalks are a common sight emerging from "Hawl cane roads." Like logging trucks in my part of the world, cane trucks rule the roads! Patches of cane can often be found in abandoned fields and along the roadside. A patch of sugar

Left:
Plate 130. *Saccharum* cultivar, sugar cane patch.

Above:
Plate 131. Sugar cane field about half grown.

cane is illustrated in Plate (130), and a field of about half grown plants in Plate (131). Short chunks of cane are available in many super markets as a snack food and cane juice can be purchased at some roadside stands.

Tournefortia—the tree heliotrope

About 100 species make up this primarily tropical genus in Boraginaceae, the borage family. *Tournefortia argentea*, the tree heliotrope, occurs on all of the main islands, except Kahoʻolawe, on several of the atolls in the northwestern group of the islands, Madagascar, tropical Australia, and extensively in tropical Asia. It occurs along the coast often just above the sandy beaches. The specific epithet "argentea" refers to the silvery appearance of the flower array as shown in Plate (132). In some sources this species is found under the names *Messerschmidia* or *Argusia* which are differentiated from *Tournefortia* by having fleshy leaves, among other features.

Orchids

Orchidaceae are one of the largest families of flowering plants with about 1,000 genera comprising perhaps 20,000 species. The family is cosmopolitan in distribution with a concentration of species in tropical regions. Many species are epiphytic. With its tropical climate and lush vegetation, one might be forgiven for expecting that the Hawaiian Islands would be home to a rich orchid flora. After all, we did see in Chapter One that the Island of Hawaiʻi, the Big Island, is often referred to as the

Plate 132. *Tournefortia argentea*, the tree heliotrope, a Pacific Basin species.

Orchid Isle. In fact, the Big Island is home to only two of the native species of orchid that occur on the archipelago. There are only three, one each in the genera *Anoectochilus*, *Liparis*, and *Platanthera*. The species representing these genera are extremely rare and localized small plants not easily found. I must confess that I have not seen any of them.

There are two orchids that one does tend to see frequently along roadsides and trails, the bamboo orchid and the Philippine ground orchid. The bamboo orchid is *Arundina graminifolia* (graminifolia = grass-like leaves), a native of southeastern

Asia and India as well as some other Pacific islands (Plate 133). It seems to be a comparative newcomer to the islands having first been collected on Oʻahu in 1945, according to the *Manual*. In somewhat wetter sites, especially along forest edges and roadsides on windward coasts, one may meet the Philippine ground orchid, *Spathoglottis plicata*. This species comes originally from southeastern Asia and was thought to have been cultivated first on Oʻahu

Plate 133. *Arundina graminifolia*, the common bamboo orchid.

in the 1920s. It is widespread in the islands now. This beautiful plant with its characteristic pleated leaves is illustrated in Plate (134).

In order to understand why Orchidaceae aren't better represented on the islands, it is necessary to consider some of the features of orchid biology. One of the criteria for ready dispersal over long distances is small seed size. At first glance, orchids would seem to fulfill this requirement admirably since they produce some of the smallest, and hence lightest, seeds of any flowering plant. A single orchid capsule may, for example, contain upwards to a million seeds. Seeds this small are

Plate 134. *Spathoglottis plicata*, the Philippine ground orchid.

categorized as "dust diaspores," propagules small enough to become part of the air column. Spores of ferns, mosses, and fungi also belong to this group, but there is a significant difference in their structure compared to that of an orchid seed. Orchid seeds owe their diminutive size to the fact that they contain only very small embryos—sometimes only a few cells—and are not yet differentiated, in contrast to more fully developed embryos in other groups of flowering plants. Orchid seeds are exceptionally vulnerable to desiccation, freezing, and ultraviolet radiation—all of which await them in the atmosphere—and they are quickly killed in seawater. Many orchids also require a close association with particular fungi which assist in acquiring nutrients for the seedling and for continued successful growth. These fungal species would normally exist in a successful orchid's home territory, but may not be present in a distant site.

A further problem involves the requirement for a specific insect species for pollination, an association that in the extreme can involve a one-on-one pairing of orchid and pollinator. There are well documented examples of certain orchid species forming a one-on-one relationship with a species of bee based upon the production by the flower and detection by the insect of a unique scent. Any disruption of these specialized adaptations can lead to extinction, possibly of both plant and insect species. An interesting tale involving this close mutualism surrounds the occasional reappearance of an "extinct" orchid species in Florida, known otherwise from one or more Caribbean islands. The newly discovered species does not last long, however, and soon disappears into extinction again. It is instructive to note that the appearance of these ephemeral species corresponds well with the occurrence of tropical storms, which obviously blew seeds, or plant parts capable of vegetative growth, the comparatively short distance to Florida, but did not happen to carry the necessary pollinator on the same flight.

Although rafting is not relevant to the transport of orchids or their seeds—recall their sensitivity to seawater and that there are no ocean currents that approach the Hawaiian Islands from the southwestern Pacific—it is useful to comment briefly on this means of propagule movement. Mats of vegetation that have been carried down tropical rivers during periods of high water have been seen hundreds of kilometers out to sea. The term "sudd" is used to identify these rafts, which I have seen making very good time down monsoon-swollen rivers in southern Thailand on their way to the sea. In addition to carrying seeds trapped in the vegetation, entire plants can travel in this manner as can insects and small animals. A highly unusual example of rafting was reported following the August 1883 eruption of Krakatau, a volcanic island lying in the Sunda Strait between Java and Sumatra. The explosion of Krakatau yielded such overwhelming amounts of pumice that rafts of this material capable of supporting the weight of a man were formed. Pumice rafts were subject to winds and currents that moved them as far east as Melanesia and to the African coast in the west, a distance of about 7,400 kilometers. The latter raft, carrying human bones, arrived on the shore of Zanzibar in July, 1884, nearly 11 months on its journey across the Indian Ocean. Other rafts were observed that carried entire, living plants, seeds, eggs, and marine animals (Zeilinga de Boer and Sanders, 2002).

A SAMPLING OF FERNS, FERN ALLIES, AND A FUNGUS

The fern flora of the Hawaiian Islands is quite extensive with many interesting species native to the islands. Island ferns range from one-cell thick filmy ferns, that inhabit only the wettest habitats such as dripping cliffs, to tree ferns that easily dwarf humans. According to Daniel Palmer (2003) in his *Hawai'i's Ferns and Fern Allies*, there are 144 native species of ferns on the islands, 106 of which occur nowhere else than the islands. Sixty-seven genera are represented on the islands, three of which are native: *Sadleria* with six, *Adenophorus* with nine, and *Diellia* with six species. *Sadleria cyatheoides* is one of the featured players in this section and is described below. Species of *Adenophorus* are primarily epiphytes in wet forests. *Diellia* is noteworthy in its flirtation with extinction, similar to a number of flowering plant genera that we have met above. Of the six known species, only one (*D. falcata*) is relatively common, but is found only on O'ahu; at least one other is thought to be extinct; and the rest can variously be described as rare or endangered.

In dealing with ferns, as opposed to flowering plants, we must focus on a totally different set of structural features. Because ferns are spore-producing plants, it is important to appreciate that the manner in which the spore-bearing bodies, the sporangia (singular: sporangium) are arranged into sori (singular: sorus). For example, sori can be located along the margins of the fronds or at varying distances inward from the margins, or along the primary veins of the fronds. A sorus can be covered with a flap of tissue, referred to as an indusium, or it can be naked. Another key feature in recognizing different ferns is the level of dissection of their fronds (leaves). Fern fronds can be simple, as in species of *Elaphoglossum* below, or more or less dissected. An example of a once divided fern (referred to as once-pinnate) is seen in *Pteris cretica*. Further dissection of fronds is seen in many ferns such as *Sphenomeris chinensis*. The combination of frond dissection, placement and nature of the sori, and overall size of plant are key features in identifying ferns.

The following examples involve ferns that are relatively easy to find and represent a variety of habitats and include Hawaiian native, indigenous, and alien species. Some of the ferns could fit as well in the next chapter where we will discuss aliens, but they are included here simply to keep all of the ferns together.

Adenophorus hymenophylloides—pai, palai huna

One of the Hawaiian names for *Adenophorus hymenophylloides* is *palai huna*, which literally means hidden fern. Although this species, which is one of nine that make up this native Hawaiian genus, is not nearly as obvious as most of the larger ferns below, a search for it will reward the interested visitor. The specimen photographed for Plate (135) was found growing on a tree on a wet, shady bank along the switchback portion of the Kīlauea Iki Trail in the Thurston Lava Tube area. This fern is a small, pendent epiphyte that can grow to a length of about 20 cm. A number of fronds clustered together give the impression of a

Plate 135. *Adenophorus hymenophylloides*, a small epiphytic fern of wet forests.

small lacy curtain. The fronds are one-pinnate with the individual pinnae only a few millimeters in length. The example in the Plate is about 15 cm long.

Adiantum capillus-veneris—'iwa'iwa, maidenhair fern

Adiantum capillus-veneris, one of the maidenhair ferns, is one of the most delicate ferns that one will find on the Hawaiian Islands (except Kaho'olawe), but the visitor must we willing to chance getting a little damp. This fern, which is not fussy as to whether it grows as an epiphyte, on rocky surfaces, or as a member of the terrestrial community, does like it wet. Palmer (2003) says that this fern is rare in the islands, but Kepler (1998) suggests that a visit to narrow gulches, deep shady valleys, or stream beds will likely get one into a probable

Top:
Plate 136. *Adiantum capillus-veneris,* a fern of very wet sites here photographed at the entrance to the Thurston Lava Tube. Photo by M. H. Hawkes.

Bottom:
Plate 137. Entrance to the Thurston Lava Tube.

habitat. It can be readily seen growing on the wet, dripping walls at the entrance to the Thurston Lava Tube where the photograph for Plate (136) was taken. The entrance to the Thurston Lava Tube viewed from just inside the tube can be seen in Plate (137). This is a great place to see many species of ferns.

Plate 138. *Angiopteris evecta* with the author shown for scale. Photo by M. H. Hawkes.

Plate 139. *Angiopteris evecta*, a broader view. The red flowers are anthuriums.

Angiopteris evecta—mule's foot fern, giant fern

Angiopteris evecta is truly a giant among ferns with fronds that can reach 20 feet in length. The stipes (main stems) of these plants can be the diameter of a grown man's upper arm as can be seen in Plate (138) with the author included for scale. Plate (139) shows a wider view of this fern which was growing in a private garden, wherein lies the problem. This fern is quite aggressive and has escaped cultivation in many areas of the tropics and is spreading and causing harm to native vegetation. In some places in the Hawaiian Islands—the mountains behind Honolulu and in the Kīpahulu Valley on Maui—*Angiopteris* has become a serious pest and can choke out native plants.

Many fern specialists think that *Angiopteris* consists of the single species *A. evecta*. Throughout its native range, however—Madagascar, tropical Asia, and the western Pacific Islands (Mabberley, 1997)—it exhibits a significant level of variation. At the far end of taxonomic opinion, however, as many as 200 "micro-species" have been described in efforts to recognize the variants in some formal manner. This species was introduced to the Hawaiian Islands in 1927 at the Lyon Arboretum in Honolulu (Palmer, 2003).

Asplenium trichomanes subsp. densa—'oāli'i

Whereas the fern we just visited above, *Angiopteris*, prefers moist, shaded valleys, our next example provides a significant contrast. *Asplenium trichomanes* subsp. *densa*, 'oāli'i to the Hawaiians, occurs on the dry, exposed ridges of Hawai'i and East Maui. Although the species is widespread in the world, the subspecies *densa* can be found only on the young mountains of those two islands where it grows abundantly. It derives its subspecific name from the dense mass of old stipes out of which new growth emerges. Plate (140) illustrates a typical clump of this fern found at an elevation of about 9,000 feet on the east flank of Mauna Kea. Plate (141) shows 'oāli'i growing near another dry-land species, *Pellaea ternifolia*, which we will meet a little further on.

Plate 140. *Asplenium trichomanes* subsp. *densa* a fern of dry sites.

Plate 141. *Asplenium trichomanes* subsp. *densa* and *Pellaea ternifolia* growing together on Mauna Kea.

Another member of this genus of ferns on the islands, characterized by very long, entire, paddle-like leaves, is *Asplenium nidus*. The leaves of this species can attain lengths of over a meter but differ in habit from other members of the genus in that the fronds emerge from the base of the plant in a fashion reminiscent of a bad-

minton shuttlecock (Palmer, 2003), albeit, an exceptionally large one. This arrangement resembles a bird's nest; *nidus* is the Latin word for nest. This fern can be found growing epiphytically in forks of trees and is widespread in the islands. Other ferns with paddle-like fronds will be met below in the discussion of species of *Elaphoglossum*.

Blechnum appendiculatum

Another example of a decorative plant that has escaped from cultivation and become a nuisance is *Blechnum appendiculatum*, illustrated in Plate (142). The photograph of this plant was taken along the switchback trail leading out of

Plate 142. *Blechnum appendiculatum*, an invasive fern. Note the normal sized fly for scale.

Plate 143. Tree ferns lining the Crater Rim Road on the Island of Hawai'i with *Metrosideros* forest in the background. Note the Datsun for scale.

Kīlauea Iki crater (Thurston Lava Tube area). Characteristic of this group of ferns is the presentation of their sori, which lie parallel to the main veins. This feature is illustrated on the frond that has been turned bottom-side up. This plant can occur in large colonies although only single individuals were seen along the trail at this site.

Plate 144. *Cibotium glaucum* showing the characteristic blue-green coloration of the species and arrangement of sori.

Cibotium—hapuʻu

This genus of tree ferns consists of perhaps eight species with representatives in southeastern Asia, Malaysia, Central America, and the Hawaiian Islands. Opinions have differed as to the number of species on the islands with Palmer opting to go with four, all of which are natives. All four species are moderately to commonly present in wet forests. Plate (143), taken along the Crater Rim Road on Hawaiʻi, shows the size attained by these tree ferns. The most common species of *Cibotium*, and one of the most attractive tree ferns, is *C. glaucum* (plate 144). The Hawaiians called this species *hāpuʻu* or *hāpuʻu pulu*. This species is easily recognized by the pale blue-green color of the undersides of its fronds, a feature clearly visible in the plate.

Plate 145. Close-up of a tree fern showing *pulu* on the stems.

Many parts of this tree fern were put to good use. The starchy core was cooked, the outer trunk fiber was used to make or line baskets for carrying plants, and the pulu was collected commercially for pillow and mattress stuffing. This product did not have much staying power, however, owing to disintegration of the fibers into a heavy granular material. Lamoureux (1976) reported that during the period 1860 to 1865 over four million pounds of *pulu* were exported. A photograph of *pulu* on the stems of a tree fern can be seen in Plate (145).

Plate 146. *Dicranopteris linearis*, the dread *uluhe*, scourge of hikers, friend to pigs.

Dicranopteris linearis—uluhe

Most people who have ventured off-trail in the Hawaiian Islands have encountered *uluhe*. It is about the most unforgiving, tangled vegetative mass one is ever likely to deal with, in the same category with catbrier (*Smilax*) or some of the blackberry species (*Rubus*). Uluhe, also known as false staghorn, is *Dicranopteris linearis*, one of about a dozen species of this tropical genus. Uluhe is an indigenous species in the Hawaiian Islands that rapidly takes advantage of disturbed areas, whether from fire, landslide, or man-made disturbances such as over-grazing by feral ungulates. It is easily recognized by its repeated branching from the terminal pair of leaves. Plate (146) shows a typical *uluhe* thicket. The observant hiker will occasionally see tunnels through the dried lower branches of this plant made by feral pigs.

Dicranopteris is a member of the family Gleicheniaceae which is represented on the Hawaiian Islands by three genera: *Dicranopteris*, which we have just seen, *Diplopterygium* and *Sticherus*. *Diplopterygium* is represented on the islands by a single native species, *D. pinnatum*. This fern bears a resemblance to *uluhe* in its branching, although it is a much larger plant. The Hawaiians recognized this size difference in their name for the fern, *uluhe lau nui*, literally, *uluhe* with big (*nui*) leaf (*lau*). It is not uncommon to find this species growing intermixed with *Dicranopteris* and *Sticherus*. This latter genus is also represented in the islands by a single, native species, *S. owhyhensis*. (Owyhee is an old spelling of Hawai'i.) This species bears a strong likeness to *Dicranopteris* which was also recognized by the Hawaiians who also called it *uluhe*. The best way to recognize the differences among these three plants is to see them side by

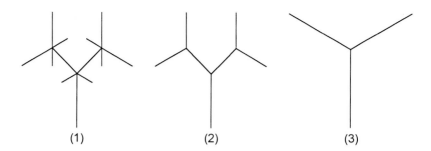

(1)   (2)   (3)

Fig. 11. Branching patterns of *Dicranopteris linearis* (*uluhe*), and related species. Patterns are: (1) *Dicranopteris*; (2) *Sticherus*; (3) *Diplopterygium*. Note that in addition to the branching pattern differences, *Diplopterygium* is a much larger plant.

side. The two *uluhes* can be distinguished on the basis of their branching—three "connections" per node in *Sticherus*, five in *Dicranopteris*. These forms are Illustrated in Figure (11). The ferns are characterized by indeterminate growth, which means that they keep producing new leaves at the ends of their branches. The terminal node of *Dicranopteris* has only three "connections" because it hasn't stopped growing.

Elaphoglossum—Stag's tongue fern

The genus *Elaphoglossum* is quite large, perhaps 600 species, and is common in pantropical and southern temperate forests. Nine species occur in the Hawaiian Islands eight of which are native. Species of this genus feature leaves that are entire and resemble, other than a stag's tongue, canoe paddle blades. In fact, the Hawaiian name for some of the island species is *hoe a Māui*, literally, Māui's paddle. Blades of some species can range up to 60 cm long. Plate (147) shows a patch of *Elaphoglossum*

Plate 147. A growth of *Elaphoglossum* on a nurse log along the Pihea Trail on Kauaʻi.

Plate 148. *Elaphoglossum crassifolium.*

Plate 149. *Elaphoglossum palaceum.*

growing on a fallen tree along the Pihea Trail on northern Kaua'i. Differences among the various species depend on the nature of the scales on the rhizomes and other more or less technical features that need not concern us.

Two species were encountered along the switchback trail on the northern wall of Kīlauea Iki Crater. Illustrated in Plate (148) is the beautiful and common native *E. crassifolium*, growing here as an epiphyte. The resemblance of this fern to a canoe paddle is also recognized in the Hawaiian name *hoe a Māui*, as above. *Elaphoglossum palaceum*, illustrated in Plate (149), is characterized by scales on its fronds, of which there are two kinds, sterile and those bearing sori. The latter fronds are the more scaly of the two types. This species, known as *māku'e* in Hawaiian referring to the brown scales on the fronds, also occurs in the Americas and parts of Macaronesia (Madeira and the Azores). This sort of unusual occurrence pattern presents a red flag to the plant geographer. Palmer (2003) suggests that this species may be a member of a group whose relationships require further study.

Filmy ferns

The so-called filmy ferns constitute the family Hymenophyllaceae whose name comes from the Greek word for membrane, "hymen," and for leaf, "phyllon." These plants are very delicate owing to the their fronds often being only a single cell-layer thick. They are particularly susceptible to desiccation and are thus found in sites that range from damp to very wet. In the Hawaiian Islands filmy ferns can be found in wet rain forests. They can be terrestrial but are more often epiphytic. The taxonomy of filmy ferns is uncertain at best. What you see in Plate (150) is a characteristic repre-

Plate 150. *Hymenophyllum* sp., a filmy fern. Photo by M. H. Hawkws.

sentative of the group, a member of what one could call *Hymenophyllum*, a name that here is taken in its broadest concept. This little plant was found growing on a tree trunk at the bottom of the stairs that lead to the Kilauea Iki trail. The area is dark and very damp and sharp eyes are needed to find the dime-sized little plants.

Grammitis tenella—kolokolo, mahinalua

This moderately large genus, perhaps 200 species in the warm tropics, is represented on the Hawaiian Islands by four species, one indigenous and three native. The plant illustrated in Plate (151) is one of the native species, G. *tenella.* "*Kolokolo*" refers to the creeping nature of this small fern while "*mahinalua*" refers specifically to this group of creeping ferns. While on the subject of word origins, it is instructive to learn that the specific epithet "*tenella*" derives from the Latin word "tenellus" in reference to the delicate nature of the fronds. This species is an epiphyte that requires a little effort to find. It occurs as tufts of

Plate 151. *Grammitis tenella.*

simple, greenish brown fronds that can be up to 20 cm long but are rarely wider than half a centimeter. *Grammitis tenella* can be easily distinguished from the other members of the genus by the positioning of its sori, which extend beyond the margins of the fronds.

Lygodium japonicum

Known as the Japanese climbing fern, this attractive little plant has gained attention for its decorative qualities. The photograph for Plate (152) was taken on the grounds of a hotel at which I stay while in Hilo. Apparently *Lygodium* has escaped from cultivation in the general vicinity of Hilo, as well as in an area near the University of Hawai'i in Honolulu, and is expanding on its own using other vegetation for support. The extent of these expansions is not known in any detail and the eventual effect on local vegetation remains unknown. The photograph shows only sterile fronds; fertile fronds carry sori on the tips of the frond lobes.

Plate 152. *Lygodium japonicum,* one of very few climbing ferns.

Nephrolepis exaltata—kupukupu

Nephrolepis exaltata, kupukupu or *'ōkupukupu* in Hawaiian, is also known as a sword fern because of its sword-like leaf shape. This is an early successional species on new lava fields where one sees it growing in cracks and often in tree wells. An example of this fern taking advantage of a tree well in the cinder field near Kīlauea Iki can be seen in Plate (28). Ferns growing in lava cracks can be found on the walk across the crater floor of Kīlauea Iki, or in almost any other suitable habitat. According to Pukui and Elbert (1986) the Hawaiian term *kupukupu* is a general term for a long, narrow fern with many lateral divisions. One of the forms of *N. exaltata*, a cultivar known as "Bostoniensis," is the well known "Boston fern."

Phymatosorus grossus—maile-scented fern, laua'e

Phymatosorus grossus, the *laua'e*, is a naturalized fern native to New Guinea, Australia, much of the South Pacific, and westward through southern Asia and tropical Africa (Palmer, 2003). It was first observed on Maui in the early 20th century and has spread widely. It is a common fern for decorative plantings, where it is frequently seen as a border plant, for quilt making, and in hula dancing. This species is also known as the *maile*-scented fern owing to the pleasant aroma given off by the leaves. *Maile* is a Hawaiian native species, *Alyxia oliviformis*, a member of Apocynaceae or dogbane family used for making *lei*. Plate (153) features *P. grossus* showing its characteristic double row of sori.

Plate 153. *Phymatosorus grossus*, the *maile*-scented fern. Note double row of sori.

Pityrogramma austroamericana—silver-back fern

Pityrogramma consists of 16 species (Mabberley, 1997; Palmer says ca. 17) with an interesting pattern of distribution including the Andes, southwestern and western North America, Africa, and Madagascar. At one time only a single species of *Pityrogramma* was thought to occur in the Hawaiian Islands and that it consisted of two forms, one with whitish waxy material, technically referred to as a farina, present on the undersides of its leaves, and one with golden-yellow material. These two forms are referred to, respectively, as silverback and goldback ferns. The treatment described by Palmer recognizes two species, *P. calomelanos* corresponding to the silverback form, and *P. austroamericana* to the goldback form. There are differences in details of leaf structure between these two species as well. Chemical studies have identified a variety of compounds in the exudates of both color forms, some of which are thought to inhibit germination and growth of potential competitors. Although this is difficult to demonstrate in Nature, experiments in the laboratory using lettuce seedlings have revealed the effectiveness of these—another example of chemical warfare in the plant kingdom. *Pityrogramma* is known as an early colonizer of fresh volcanic fields in Mexico. Plate (154) shows a specimen photographed along the Kukui Trail on Kaua'i. I have also seen it growing along the northern part of the Power Line Trail (road), also on Kaua'i, and in a variety of other arid areas as well. Recent studies by fern specialists suggested that these species should be placed in their own genus, *Pentagramma*.

Plate 154. *Pityrogramma austroamericana,* the gold-back fern.

Pellaea ternifolia—*kalamoho lau liʻi, kalamoho*

This little fern, *Pellaea ternifolia,* commonly known as cliff break, can be found in the same habitats as *Asplenium trichomanes* subsp. *densa* that we met a few paragraphs earlier. Both of these ferns thrive well

Plate 155. *Pellaea ternifolia* growing on the wall of an old lava sink on the Mauna Loa Trail.

on dry, exposed sites on all of the major islands, but they are much more frequently met on Hawaiʻi and east Maui. Plate (155) shows a plant growing on the face of an old lava sink along the Mauna Loa Trail. Plate (141) shows the two dry site species growing together on Mauna Kea. This species also occurs in North and Central America (including Mexico) and on other Pacific islands.

Polypodium—*ʻae, ʻae lau nui*

The genus *Polypodium* (polypody ferns; literally, many footed) consists of about 150 species worldwide with a generous representation in the tropics. *Polypodium glycyrrhiza,* the licorice fern, is a well known North American species whose rhizomes can be eaten. A most peculiar feature of *P. vulgare* is the presence in its rhizomes of

Plate 156. *Polypodium pellucidum* var. *vulcanicum,* often seen growing with *Stereocaulon vulcani,* the white lichen.

exceptionally high concentrations of a hormone originally isolated from silkworms. *Polypodium pellucidum* is the Hawaiian native member of this genus. Three varieties are recognized, one of which, var. *vulcanicum,* as its names suggests, can be found growing on open lava fields. It is shown in Plate (156) along with the lichen *Stereocaulon vulcani.* This photograph was taken at about 5,500' elevation along the Saddle Road on Hawai'i. This fern can also be seen in abundance in Haleakalā National Park.

Pteris cretica—'ōali

Pteris consists of about 250 species in tropical and warmer temperate areas of the world and is represented by three native, two indigenous, and one naturalized species in the Hawaiian Islands. Our example here is *P. cretica,* one of the indigenous species. The photograph for Plate (157) was taken along the Kīlauea Iki Trail at the Thurston Lava Tube end of the trail, but occurs in dry to

Plate 157. *Pteris cretica.*

mesic areas on all of the major islands. In maturing plants the lower pair of pinnae are divided into nearly equal branches. Young individuals will not exhibit this feature.

Plate 158. *Sadleria cyatheoides,* note characteristic young red leaves.

Sadleria—'amau'u

The genus *Sadleria*, which consists of two to six species depending upon authority (Palmer says six), is native to the Hawaiian Islands. The species most easily seen in Nature is *S. cyatheoides*, known to the Hawaiians as *'ama'u*. A very distinctive feature of this plant is the salmon colored young leaves (Plate 158). As the leaves age the red coloration fades and the leaves take on a shiny dark appearance. The red pigment may serve as a barrier to the strong sunlight that this plant experiences in its exposed habitats, commonly on open lava fields. It is readily found in the vicinity of Kilauea Volcano;

Plate 159. *Sadleria squarrosa,* a smaller species of *Sadleria.*

in fact, the name of the fire pit in Kīlauea Volcano, Halemaʻumaʻu, literally means "home of the ʻamaʻu." One of the more interesting places to view this fern is on the Halemauʻu Trail in the vicinity of Koʻolau Gap where the trail begins to descend; this is a particularly striking area with mist clouds pouring in from the northeast. One can also find this fern along Route 200, the Saddle Road on Hawaiʻi and frequently in wet forests as well. A red dye was prepared from young leaves of this fern and the *pulu*, the soft hairs at the base of its leaves, was used as a packing material for pillows. *Pulu* was also used occasionally as an absorbent in the preparation of a body for burial or cremation. A somewhat smaller species is *S. squarrosa* (Plate 159) which can be seen growing on moist, shaded banks. The pictured specimen was also found on the northern switchback portion of the Kīlauea Iki Trail (Thurston Lava Tube end).

Sphenomeris chinensis

Sphenomeris chinensis is one of the most common ferns on the Hawaiian Islands occurring in a variety of habitats, including mesic to wet forests, grasslands, and along streams. It is also a widespread species in Madagascar, Himalayas, India, Sri Lanka, Malay Peninsula, Japan, Taiwan, the Philippines, Polynesia, and China, from which it derives its specific epithet (Palmer, 2003). This species owes its common name, the lace fern, to its deeply cut leaves (Plate 160). The Hawaiians have several words for the lace fern, including *palaʻā*, *palapalaʻā*, and *pāʻū o Palaʻe*. Leaves of the lace fern are used to make lei and a brown-red dye can be prepared from its stems (Krauss, 1993). The fern was also used to decorate the altar in houses that were used to teach *hula* and are still used in *hula* decorations today.

Plate 160. *Sphenomeris chinensis,* the very common lace fern.

Lycopodiella

Lycopodiella is a small genus, perhaps as many as 15 pantropical species, related to the much larger and perhaps more familiar *Lycopodium*. A single species of *Lycopodiella*, *L. cernua*, grows widely in the Hawaiian Islands and can be found in wet habitats as well as dry ridges and sites in between. In Plate (161) it is seen growing amongst the *uluhe*, *Dicranopteris linearis*. The Hawaiians referred to this species as *wāwae'iole*, meaning rat's foot, or *hulu 'iole*, meaning rat's hair. Specimens of this attractive plant can resemble small Christmas trees.

Plate 161. *Lycopodiella cernua*, one of the largest fern allies on the islands.

Psilotum nudum—moa

This ancient genus consists of two species, *P. nudum* (Plate 162) and *P. complanatum*, both of which occur in the Hawaiian Islands. These plants can be terrestrial or epiphytic and occur widely in tropical regions of the world. The two can be distinguished by their stems, which are triangular in cross section and upright in *P. nudum*, giving rise to its common name of "upright whisk fern," and drooping in the case of *P. complanatum*, which has flat (planar) stems. Reproductive structures, the sporangia, or spore-bearing organs, occur as small ball-like structures distributed along the stems of both species. The Hawaiian name for these species is *moa* or *moa nahele*, among others. *Moa* is the Hawaiian word for chicken—*moa nahele* literally is the forest *moa*.

Plate 162. *Psilotum nudum,* the so-called whisk fern.

AND A FUNGUS

One of the most attractive fungi that a visitor to the islands might find is illustrated in Plate (163). This organism, whose name is *Aseroe rubra,* can be found among leaf litter in variety of habitats, but it prefers fairly damp sites. I have seen it along the Awaʻawapuhi Trail in northwestern Kauaʻi, in the Limahuli Garden on the north coast of Kauaʻi, and along the western access trail to the Man in the Mountain on the eastern coast of Kauaʻi near the town of Kapaa. The species occurs widely in tropical Pacific Islands.

Plate 163. *Aseroe rubra,* a fungus that grows in damp, rotting leaf litter.

Alien Invasion

I had chosen the title for this chapter before I learned of the existence of Robert Devine's 1998 book of the same, but expanded, name, *Alien Invasion. America's Battle with Non-native Animals and Plants*. It so perfectly describes the situation that it is not surprising that others also see the phenomenon as an invasion of loathsome and frightening creatures. Although we are not dealing with "things from outer space," but instead are often concerned with majestic trees and beautiful flowers, the image may seem a bit overblown. In reference to an island ecosystem, however—isolated by both time and distance—even the most beautiful flower has the potential to become both loathsome and frightening, as we will see in the case of the "green cancer" below. Devine includes mainland and island invaders, and shares his personal experiences with individuals who have immediate involvement with the plants and animals concerned. Read his book to appreciate the full impact that unwanted visitors can have; do not expect a happy ending.

In this chapter we will meet several plant species whose presence in the Hawaiian Islands represents a serious problem. Most of the troublesome species are present on the islands through human activity—both intentional and careless. A few indigenous species can be serious pests as well, but in most of these cases, human activity can be seen as a contributing factor to their invasive success. Since about half of the species listed in the *Manual* are described as "naturalized," we can do little more here than provide a sampling, as was the case in Chapter Two, where we met only a few dozen native species. By focusing on a representative sample of invasive species, rather than simply documenting dozens of offenders, it is possible to get a closer, more personal view of the problem. Most of the species chosen for this chapter are easily found in the wild, far too easily for many of them. It is hoped that the visitor to the islands can take away a feeling—not necessarily of panic—but certainly one of informed concern about the impact these organisms can have, and indeed are having, on an island ecosystem and what efforts are needed to prevent further invasions and limit the damage already occurring.

To start with, it is useful to define what we mean by a naturalized species. In addition to having arrived on the islands through human activity, the species must have become thoroughly established and be capable of reproduction either vegetatively or sexually in its new habitat. As the reader may recall from the introduction, the current flora of the Hawaiian Islands consists of about 50% naturalized species. Not all of these are noxious weeds or offer a particular nasty threat, but enough troublesome species exist in the islands to make a complete survey of them too long a topic for this book. The species that make up this chapter are ones about which enough is known to provide the reader with an idea of the challenges faced by the native species themselves, as well as by those workers who are trying to eradicate the intruders. Four specific sources of information were used for preparation of this chapter. The most detailed of these is *Alteration of Native Hawaiian Vegetation: Effects of Humans, their Activities and Introduction* written by L. W. Cuddihy and C. P. Stone (1990). In a similar vein, *Conservation Biology in Hawai'i*, a collection of essays edited by C. P. and D. B. Stone (1989), presents an overview of the biological, philosophical, and sociological aspects of conservation of island systems. A concise coverage of invasive species, covering plants and animals, including color photographs, is *Hawai'i's Invasive Species* by G. W. Staples and R. H. Cowie. *Plant Invaders. The Threat to Natural Ecosystems* by Q. C. B. Cronk and J. L. Fuller (2001) treats the topic of invasive species in broader scope but includes some examples from the Hawaiian Islands.

Examples will be presented in two batches, the first featuring some of the worst actors on the islands, and the second plants that, while clearly alien, do not present nearly the problems caused by the first group. I include them because they represent groups of plants that may be more or less readily recognized by visitors from the northern temperate areas. Most visitors are not likely to be familiar with many tropical species but are certainly familiar with plants that they might see in their own gardens. We start with some of the nastiest of all, the melastomes featuring their chief evil-doer, *Miconia calvescens*, the "green cancer," and some of its conspiratorial relatives.

Melastomes—the curse, green cancer, and others

Melastomataceae are a moderately large family consisting of nearly 5,000 species in 188 genera (Mabberley, 1997). The family is heavily represented in South America, but does occur naturally in other warm and tropical regions of the world. Identification of a member of this family is quite easy by noticing the typical venation pattern of the leaves, which is illustrated in Plate (164). The Hawaiian Islands have their fair share, 14 naturalized species in 11 genera. There are no native melastomes in the islands. This family is home to some of the most aggressive plants that one is ever likely to encounter. The problem, as usual, is that some of them are quite attractive; others just seem to travel well; while most of them seem to be particularly capable of becoming established just about anywhere. One of the more attractive members of the family resident on the Hawaiian Islands, illustrated in Plate (165), is *Tibouchina urvilleana*, known commonly by various names including lasiandra, princess flower, and glory bush. This species is native to southern Brazil and has been widely cultivated because of its attractive flowers; I keep several as house plants. Extensive growths of this species occur along the lower third of Highway 11 that runs from Hilo to the Hawai'i Volcanoes National Park. Authors of the *Manual* inform us that it was first collected on Hawai'i in 1917 near the little town of Kurtistown through which one drives on Highway 11 on the way to the national park.

Plate 164. Characteristic leaf venation pattern of the family Melastomataceae (melastomes).

Plate 165. *Tibouchina urvilleana*, lasiandra, princess flower, or glory bush, a pest by any name.

Plate 166. *Clidemia hirta*, Koster's curse, worse pest than *Tibouchina*.

Several other melastomes also distinguish themselves as quite serious pests. Among these is *Clidemia hirta* (Plate 166) known as "Koster's curse." The photograph for the plate was taken near the southern trailhead of the "Powerline Trail" on Kaua'i where the species was growing abundantly beside the trail. This aggressive weedy shrub is a native of Central and South America but has become extremely widespread in tropical regions of the world, including the Hawaiian Islands, Fiji, Java, Samoa, British Solomon Islands, Tonga, Palau, Madagascar, Tanzania, Sri Lanka, India, Singapore, and Sabah (Cronk and Fuller,

2001). The routes of dispersal of this plant remain unclear, but it certainly would seem to have been accidental since the plant has little to offer either aesthetically or commercially. Cronk and Fuller go on to point out that some 40,000 hectares on Oʻahu alone are infested. This is a problem of gargantuan magnitude. Could anything have been done to avert this disaster? Unless one could foretell the future, likely not. It is interesting to note that the plant had been known in both the Hawaiian Islands and Fiji for some 30 years before it became clear that it was a serious source of trouble!

Can anything be done in circumstances like these to eradicate the pest or at least slow its spread down a bit? Physical removal of a plant this widespread and well established is virtually impossible. For one thing, there is likely to be a seed load in the soil that will last for years if not decades. Wholesale application of herbicides, while certainly feasible, runs the very considerable risk of killing everything else in the vicinity as well. In situations such as this, the ideal remedy lies in utilizing some biological control organism, something that would likely have kept the species in check in its native habitat. One attempt at biological control, which was very successful in one situation, did not work in another. A species of host-specific thrips from Trinidad was introduced into Fiji in 1930 and released into the field. Within a few months the insect, which attacks terminal shoots, had brought the *Clidemia* population under control. The insect was introduced to the Hawaiian Islands in 1953, but the results were discouraging. The insect works best in full to near full sunlight and thus had no effect upon *Clidemia* plants in shaded habitats. The insect was also preyed upon by local Hawaiian insects. Introduction of a moth from Puerto Rico that attacks *Clidemia* has also been attempted but has not yet been effective. Another approach has been to utilize pathogenic fungi as control agents. A fungal pathogen isolated from *Clidemia* in Panama is undergoing tests in the Hawaiian Islands (Cronk and Fuller, 2001).

These examples are typical of the biological control approach to attacking unwanted guests. Although the approach sounds as though it ought to work, especially when dealing with a host-specific insect or fungus, it is important to bear in mind the likelihood that the entire ecological background in which the control organism normally functions (the fungus) has likely been altered. This was clearly the case with sensitivity to light, or rather its absence, in the case of imported thrips mentioned above. Whereas the invasive plant may be able to survive under somewhat broadened ecological conditions in its new habitat, perhaps the control agent does not possess the same wide tolerance. Each situation has to be tested and evaluated in the field. All of this takes time and a good deal of resources.

Plate 167. *Miconia calvescens,* the nastiest of all. Note characteristic purple under side of leaf and the size of the leaves in such a young plant.

Robert Hobdy, quoted in Robert Devine's *Alien Invasion* book, says that as far as invasive species are concerned, *Miconia calvescens* is "…in a class by itself." Even more stark was the statement made by Dr. R. Fosberg, notable Hawaiian botanist, that all by itself this plant has the capability of wiping out Hawai'i's entire native flora! *Miconia calvescens* is native to West Africa but has found its way to many new homes including Sri Lanka and Tahiti where it has become naturalized. It escaped from a botanical garden in Tahiti and quickly became naturalized and is widespread on the island where it is known as the green cancer. Its presence in the Hawaiian Islands dates to 1961 when a specimen was acquired by a botanical garden on O'ahu. A warning in 1971 that this plant could bring serious problems to Hawaiian forests went unheeded, and by 1992 it became clear that the prophecy was correct. Because of its large leaves, attractively pigmented on their undersides (Plate 167), this species became a popular decorative plant for private gardens. Unfortunately, it has escaped from cultivation and has been spreading. A healthy growth of this species can be seen along the steep banks that line the Pepe'ekeo Scenic Drive north of Hilo in the vicinity of the Hawai'i Tropical Botanical Garden.

In order to help monitor the appearance and spread of this noxious pest, the Department of Land and Natural Resources (of the State of Hawai'i) has circulated a **Wanted: Miconia—Dead or Alive** poster listing telephone hotline numbers on the major islands. Anyone finding this species of *Miconia* in the wild is asked to report the sighting to the authorities for verification. For sightings on the Island of Hawai'i one is encouraged to call the Weed Control Specialists: in Hilo at 933-4447; in Kona at 323-2608; on Moloka'i at 567-6150; on O'ahu at 973-9538; or on Kaua'i at 241-

Plate 168. *Passiflora mollissima*, banana poka.

3413. National Park Service representatives can be contacted at 572-1983 at Haleakalā (Maui), and at 967-8211 at Volcano (Hawai'i).

Banana poka

The word banana means what it says; the word "*poka*" in Hawaiian is the same as the word "*moka*," whose meanings include offal, waste matter, refuse, and filth. Thus the common name of *Passiflora mollissima*, filthy banana, more or less, suggests something not quite good. This member of the passion fruit family, Passifloraceae, is one of the most aggressive weedy species in the islands. It is a vine with the capacity to attach to, climb on, and crush through shear mass almost any forest tree it encounters. It has been likened to the infamous kudzu vine (*Pueraria montana* var. *lobata*), one of the most aggressive weeds of the Atlantic States. (For those readers not familiar with kudzu, it was introduced into eastern North America from northern Asia for erosion control and to produce a green fodder. It is capable of very rapid growth. A colleague of mine at the University of Georgia claims that he can hear it growing at night.) Banana *poka* is native to South America where it is cultivated for its fruit, but its introduction to Hawai'i, to cover outbuildings, according to all reports, has not been celebrated. It is a serious pest in the Koke'e area (Kaua'i) and on Hawai'i where it grows in mesic forests. Plate (168) illustrates this species.

     An interesting way to look at the impact of alien species on an island is to consider the combination of invasive plants and natural phenomena, my favorite being the tropical storm. Tropical storms are naturally occurring events in the Hawaiian Islands, as they are in all of the Pacific Basin. In extreme cases they may reach hurricane proportions as witness the devastating visit by Iniki in 1992. Forests can be heavily damaged by the high winds—effects of Iniki on Kaua'i's forests are still visible over 10 years later—and valleys can be devastated by flooding. Blow-downs and flooding leave behind expanses of land, either badly disturbed or completely devoid of vegetation. Before Human colonization of the islands, these areas would in time have regenerated by vegetative growth or from seeds. With the importation of alien plants the situation changed drastically. Many of the newcomers were very adept at taking advantage of newly opened sites in forests or in river valleys or much of anywhere else. Some of the newcomers were fast-growing species—not uncommon for weeds—and soon became established, out-competing native species for nutrients or water, or over-topping natives species and depriving them of light. Once the newcomers had become entrenched, there was essentially no way for the native plants to become re-established. Banana *poka* is a good example of an opportunistic,

fast growing species, as are the melastomes, the gingers, the guavas, the firetree, and an assortment of alien grasses, to name just a few. We will look at some of these below.

Myrica faya—the fire tree

If a person were asked to design the perfect pest, it would be difficult to do better (or worse?) than the firetree, a fruiting branch of which is shown in Plate (169). The firetree is *Myrica faya*, a member of Myricaceae, a small family of only three genera, two of which have only one species each. *Myrica* would likely be known to North Americans as the wax myrtle or scented bayberry (M. *cerifera*) in

Plate 169. *Myrica faya*, the firetree.

the eastern states, and sweet gale or bog myrtle (M. *gale*) in the western states. *Myrica faya* is native to the Macaronesian Islands (Azores, Canary Islands, and Madeira) and has become naturalized in southern Portugal. It was Portuguese colonists who brought M. *faya* to the Hawaiian Islands in the 1880s for decorative plantings and to use the fruit for wine-making. A brief history of the tree's spread in Hawai'i Volcanoes National Park shows the species' capacity to replace native plants. One tree was known to exist in the Park in 1961 near the Kīlauea Military Camp (close to Volcano House), although there may have been others in the forest behind the camp. By 1966 firetree was present on 90 hectares (225 acres) in the Park. By 1977 the number had grown to 3,640 ha (9,000 acres) and by 1985 to over 12,200 ha (30,130 acres). Across the entire state the area involved by the mid 1980s was 34,000 hectares, nearly 84,000 acres! What makes this species so successful?

One of the significant factors in explaining the expansion of firetree is its capacity to provide its own nitrogenous building blocks. Associated with the roots of this species is a bacterium with the capacity to fix atmospheric nitrogen. This capacity provides essential nutrients to the tree allowing it to survive in marginal habitats. It has been estimated that this species returns to the earth at least four times the amount of nitrogenous material than all other native vegetation in a comparable area. It is, in a word, a portable fertilizer factory. The addition of that much nitrogen to the environment upsets the nutrient balance in such a way that native species cannot compete. It is also a very fast-growing species that likely out-competes native species for water. A high seed yield coupled with a high rate of germination provides overwhelming numbers of recruits. Movement of seed is accomplished with the help of alien birds, the Japanese white-eye among others, and feral pigs who like to eat the fruit. In addition to distributing seeds at some dis-tance from the source, each of these vectors deposits a little of its own fertilizer in the process. Firetree is also capable of suckering, a process that can lead to thick-ets of trees in a short time. If all of these characteristics weren't enough to make this species nearly unstoppable, there has been speculation that it is also capable of allelopathy, the capacity to produce and deposit chemicals in the soil that inhib-it seed germination of possible competitors. Details on this aspect of the firetree are lacking but related species have been shown to engage in this form of chemical warfare.

Removal of pest species this well equipped for survival presents nearly insur-mountable challenges. Physical removal is labor intensive and can be very expensive, although volunteer efforts have made some progress in a few places. Evidence of cut-ting can be seen in the Devastation Trail area of Hawai'i Volcanoes National Park, but continued efforts are necessary owing to the seed load in the soil. Use of chemi-cal deterrents runs the unacceptable risk of doing as much harm to native vegetation as it does to the pest. Repeated spraying using selected herbicides has had some effect on reducing the number of trees in some sites but this is a costly approach and entails hard work. The seeds will germinate almost anywhere they fall or are deposited including places difficult to get to by workers trying to eradicate the plant.

Biological control is always a potential remedy, but here the risk of introduc-ing yet another foreign organism into the environment comes into play, a risk not to be taken lightly. Considerable study is necessary to be certain that this approach will work. During these activities, unfortunately, the tree continues to expand its range.

Drastic problems often require drastic solutions. It occurred to me that a possible way to attack the problem might be to devise a way to eliminate, weaken, or at least reduce the symbiotic relationship between the tree and the nitrogen-fixing bacterium. Since the stepwise establishment of the host-parasite relationship is very well known, it would seem possible, in theory at least, to interfere with one of the steps involved in the association using a genetically modified bacterium. Inoculation of soil in the vicinity of young firetree saplings with a modified bacterium might interfere with the establishment of the association to sufficient extent to compete with the natural bacterium. Even if only a fraction of 'normal' associations were established, perhaps the competitiveness of the tree might be reduced. Unfortunately, the use of genetically modified organisms in Nature presents its own array of problems, not the least of which is political. Although this approach might have potential to help solve the problem, it could well bring with it a measure of controversy.

Lantana camara

Another popular decorative plant familiar to visitors from the mainland is *Lantana camara*, often referred to simply as lantana. This member of Verbenaceae (verbena family) is native to the West Indies but has been extensively naturalized worldwide. Lantana is a popular flowering shrub seen in many plantings in southern California where it can be purchased from many garden shops. The flower head presents an attractive array of yellow individual flowers that turn to pink when pollinated. The leaves are sticky and have a somewhat unpleasant odor. This species is a serious pest in open places, pastures in particular, where dense thickets form. Range animals generally avoid it; it is poisonous to some. Most parts of the plant are poisonous; the fruits are eaten by birds, however, which accounts for its widespread distribution. Although the plant can be found along roadsides and in waste land almost anywhere, one of the most impressive dis-

Plate 170. *Lantana camara*.

plays that I have seen occurs on the hillside pasture near the beginning of the Boy Scout Trail on Maui, which is where the photograph for Plate (170) was taken.

Leucaena leucocephala

Leucaena leucocephala, a member of Fabaceae (Leguminosae), was introduced to the islands in the 1830s with the idea that it would serve as both cattle feed and as a source of firewood. This species exhibits the typical capacity of legumes to form a symbiotic relationship with nitrogen fixing bacteria so that it too is well equipped to colonize marginal land. This plant is so widespread in the islands it seems almost unnecessary to give directions to find it! However, to be consistent, I can suggest looking along the roads leading to Koke'e Park at lower elevations along Highway 550. There is also a healthy growth along the main road (Highway 520) leading to Poipu Beach in southern Kaua'i. Plate (171) illustrates the small puffs of white flowers of the plant, hence the specific epithet *leucocephala* (literally, white head), while Plate (172) shows the characteristic seed pods. In the Hawaiian language this plant is known as *koa haole*, literally foreign *koa*.

Plate 171. *Leucaena leucophylla,* flowers of *koa haole.*

Plate 172. *Leucaena leucophylla,* characteristic seed pods of *koa haole.*

Koa haole is a prolific grower and can form monotypic stands excluding all native plants. It also produces large seed crops resulting in a seed load in the soil sufficient to renew the crop for many years to come, even if all the mature trees were removed from an area. Feral animals have been somewhat effective in keeping growth under control but removal of the animals resulted in re-establishment of the *koa haole* thickets. Some indication of the apparent ecological perversity of this plant, although it doubtless isn't unique to it, is a comparison of its existence in the Hawaiian Islands with its existence in Martinique. Over a period of years the native vegetation of this Caribbean Island slowly regained dominance with the eventual elimination of *Leucaena*. Over a period of 40 years there is no evidence in the Hawaiian Islands that native species are making any gains against *koa haole*. This is a strong reminder that an intimate knowledge of local ecological factors is vital in attempts to combat invasive pests.

Perhaps a good word should be added to indicate that the genus *Leucaena* is not intrinsically bad. Its species have many uses including livestock food, green manure, a potential oil source (seeds) for energy production, soil stabilization and conservation, and with some species, an edible seed crop (some species are poisonous, however). There are *Leucaena* improvement breeding programs underway at the present time in several countries. As is obvious, and this is undoubtedly true for all aliens that we will meet in this book, it is only when these species are introduced to places where they really don't belong that serious problems arise.

Plate 173. *Melinis minutiflora,* a patch of molasses grass.

Plate 174. *Melinis minutiflora* individual.

Melinis—molasses grass

I have chosen molasses grass as representative of invasive grasses because, it is sad to say, it is very easy to find. Its proper name is *Melinis minutiflora* and it is a native of Africa. This grass is widespread in the tropics where it has been introduced as a fodder grass. It occupies dry to mesic sites, forms a dense ground cover which prevents native seedlings from becoming established, and is fire adapted. This invasive species brings the serious threat of intense fires to dry forests, thus putting the majority of native species at risk because they are not protected in any way. Molasses grass occurs on all of the main islands except Ni'ihau. Plate (173) illustrates a patch of molasses grass growing along the Boy Scout Trail; Plate

(174) shows an individual plant. This plant's common name is derived from the sticky material on its leaves that smells very much like molasses. Hiking boots, socks, and trouser legs easily become covered with this sticky material. One of the results of doing field work in drier parts of the islands is the combination of molasses grass and dust from weathered lava. Eradication of this species presents a problem because it does provide some fodder for range animals and its complex root system does bind soil. The question posed by plants such as molasses grass is simply whether their virtues outweigh its negative features? In this case, particularly with regard to the potential harm caused by fire, probably not.

Guava

The genus *Psidium*, to which the guavas belong, is a member of Myrtaceae, the family that contains *Metrosideros*, which we saw in Chapter Two, and the familiar *Eucalyptus*. Two species of guava have become naturalized in the Hawaiian Islands, *P. guajava* and *P. cattleianum*. Although both species have their virtues, as a source of fruit for jams, jellies, and a variety of fruit juice preparations, they nonetheless tend to be aggressive weeds, especially *P. cattleianum*. Both species occur widely distributed on the islands in places forming extremely dense thickets to which the term "dog hair forest" has been applied. *Psidium cattleianum*, the strawberry guava, 'ula'ula, is considered to be one of the most serious weed problems on the islands. Its fruits are attractive to a variety of animals, birds and pigs in particular, with resultant seed dispersal by air and by ground. Strawberry guava prevents re-establishment of native

Plate 175. *Psidium cattleianum*, a "dog-hair" patch of strawberry guava. It is possible that this is a single colony.

vegetation through allelopathic suppression of growth as well as through nearly total blockage of sunlight owing to the density of the thickets. An example of such a thicket lies on the Boy Scout Trail, here shown in Plate (175). Another excellent example of a strawberry guava forest can be seen on the right hand side of the highway as one drives along Highway 11 from Hilo to the Hawaii Volcanoes National Park between about the 1,000' and 2,000' elevation markers. One of the few things that seems able to grow with the strawberry guava, at least at its margins, is *Tibouchina*, which belongs to another group of highly invasive weeds, the melastomes, which we already met above.

Grevillea—the silk oak

Proteaceae provide some of the most spectacularly beautiful flowers known in the horticultural trade, species of *Protea* likely being the most familiar to most readers. Hardly a visitor to the Hawaiian Islands misses the opportunity to try another of this family's members, the macadamia nut which is the edible seed of members of the genus *Macadamia*. Another attractive member of this family is the Australian native *Grivellea robusta*, the silk oak. Silk oak has been widely planted in the islands both for its attractive flowers, illustrated in Plate (176), and as a source of timber. Over two million trees were planted in the islands between 1919 and 1959 (except on Kaho'olawe) for timber production according to the *Manual*. The problem with this species is its capacity to withstand draught, which obviously gives it an advantage

Plate 176. *Grivellea robusta*, the silk oak.

over local species that are not so tolerant, and its capacity to expand into both disturbed and natural areas as well as areas with very poor soil cover.

Spathodea campanulata — African tulip tree

Without a doubt *Spathodia campanulata* (syn. *S. nilotica*) is one of the most strikingly beautiful flowering

Plate 177. *Spathodea campanulata,* the African tulip tree.

trees that one is likely to see in the Hawaiian Islands. Its common names, African tulip tree, fire-bell tree, flame of the forest, flame tree, fountain tree, and Nandi flame, hint at its most attractive feature. This species, the only one in the genus (the family is Bignoniaceae), is a native of tropical Africa but it has been widely planted in tropical and subtropical areas in other parts of the world as a decorative tree. This species can become a weedy pest in low elevation forests where its seedlings, because of their shade tolerance, compete well with native plants. Its seeds are equipped for aerial flotation and can be dispersed widely by this means. Flame trees can be found widely in many cities as well as in more rural areas in the islands. Plate (177), taken near Poipu Shopping Mall, illustrates the distinctive floral array of the flame tree.

Gingers

The gingers, members of family Zingiberaceae, are well represented in the Hawaiian Islands, although there are no endemic members of this widespread tropical family. The common ginger of the kitchen is *Zingiber officinalis* whose rhizome can be used as is, or dried and powdered, or preserved in a syrup. The *Manual* tells us that there are eight naturalized species representing five genera of the family in the islands, although other members of some of these genera are under local cultivation. Two of the species were originally brought by Polynesian settlers, *Curcuma longa* and *Zingiber zerumbet. Curcuma longa* is the common turmeric used in the preparation of curries and for dying silk and wool. The Hawaiians, who called the plant 'ōlena, apparently used it as a source of dye for *tapa* cloth. Kepler (1993) notes that leaves were used to flavor meats and fish. The mature flower heads of *Zingiber zerumbet*, called 'awapuhi or 'awapuhi kuahiwi in Hawaiian, or simply shampoo ginger in English, exude a slimy, aromatic sap that was used to scent tapa cloth and as a shampoo. Neither of these specific Polynesian imports has escaped to become a nuisance. The same cannot be said for the next plant.

Hedychium gardnerianum, known as *kāhili* ginger, or just *kāhili*, is a native of the Himalayas. It is thought to have entered the islands in the early 1900s probably intentionally as an ornamental. It flowers are quite beautiful (Plate 178) and produce a powerful and very pleasant aroma. A flowering *kāhili* ginger can often be smelled before it is seen. This plant was first collected in Hawai'i Volcanoes National Park in 1940 and has since become a serious pest in the park and elsewhere. It frequently forms a homogeneous grove and in so doing totally excludes local species. It is highly shade tolerant and virtually blankets the ground. Its red fruits are attractive to birds who help spread the plant to new locations. *Kāhili* ginger, unfortunately, is all too easy to locate in the field. The southern end of the Kilauea Iki Trail (Hawai'i Volcanoes National Park) passes through a large growth of the plant, possibly a football field in area. This species has also become well established in the vicinity of Koke'e Park on Kaua'i including along parts of the Awa'awapuhi Trail where the photograph for the plate was taken.

Plate 178. *Hedychium gardnerianum*, the Kāhili ginger.

A somewhat less aggressive ginger is *Hedychium coronarium*, which travels under several aliases, *'awapuhi ke'oke'o* in Hawaiian, and common ginger lily, butterfly lily, or garland flower in English (*Manual*). It is thought to have been introduced as a decorative plant and escaped from cultivation late in the 19th century. It is native to southeastern Asia and is much used for decoration and is well known to have become naturalized in many

Plate 179. *Hedychium coronarium*, the crown or white ginger.

tropical countries. The plant pictured in Plate (179) was found growing along the main highway south of Hilo where it occurs abundantly in wet ditches.

Rubus—the blackberries

Authors of the *Manual* point out that their definition of species, with regard to the five non-natives, is a very broad one. Species definition in the blackberries can be exceedingly difficult owing to their capacity to undergo self fertilization, as well as a strong tendency to form hybrids. The Prickly Florida blackberry (*R. argutus*) is a native of central and eastern North America and has become an extremely serious weed in a range of disturbed habitats on all of the Hawaiian Islands, where it was first collected in 1904. Its capacity to thrive in a variety of habitats is one of the features that typifies invasive species. This invader can be found in some abundance along the Pihea Trail a bit beyond the Kalalau lookout. It can also be found growing along the Devastation Trail in Hawai'i Volcanoes

Plate 180. *Rubus argutus.*

National Park where the photograph for Plate (180) was taken. A more recent immigrant is the Yellow Himalayan blackberry (*R. ellipticus*), which has become naturalized in at least two areas on Hawai'i since its first report on the island in 1961. This plant is native to tropical and subtropical India but is grown in other parts of the world in suitable habitats. *Rubus sieboldii*, whose home range includes southern China, Japan, and Okinawa, was first noticed in Kaua'i in 1970 where it seems to be doing well. Another immigrant from southeast Asia is the Hill or Mysore raspberry (*R. niveus*), which has been cultivated in the Hawaiian Islands for many years and has become naturalized in sites on Maui and on Hawai'i. Of longer residence is the Thimbleberry or Mauritius raspberry (*R. rosifolius*), which arrived on the Hawaiian Islands in the 1880s and has become established as a common weed.

The next group of aliens are not considered to be dangerously aggressive, at least not to the extent that some in the first list are. Nonetheless the species listed have made a home in the islands and appear to be capable of spreading. Whether they ever become threatening only time will tell. Many other aliens could be listed,

Plate 181. *Anemone hupehensis,* a large visitor from eastern Asia.

Plate 182. *Buddleia asiatica,* dog tail to the Hawaiians, growing on the crater floor of Kīlauea Iki.

Plate 183. *Buddleia asiatica* in fruit, growing along the Saddle Road.

but the ones included represent groups of plants that most visitors will recognize as natural members of North American gardens.

Anemone

A single species of *Anemone,* a common genus within Ranunculaceae, occurs in the Hawaiian Islands. *Anemone hupehensis* is an Asian species that has escaped from cultivation and finds suitable habitats in wasteland, along roadsides, and also in Hawai'i Volcanoes National Park. The specimen photographed for Plate (181) was growing, with several others, in the open cinder fields along the Devastation Trail boardwalk in the Park. This anemone can attain an impressive size—a meter and a half—according to authors of the *Manual.*

Buddleia asiatica

Buddleia (sometimes spelled *Buddleja*) *asiatica,* a member of Loganiaceae, is a shrub or small tree that features a spike of small white flowers. From this floral display it is easy to see how the Hawaiian name *huelo 'ilio,* literally dog tail, came into usage. The photograph for Plate (182) was taken on the floor of Kilauea Iki where the plant was found growing in a crack in the lava. This is a typical location for a species that does well in disturbed areas, although it might be listed as a pioneer species by some. Plate (183) shows a mature plant in seed photographed along the Saddle Road on the Big Island at about 5,000' elevation. The seed yield in these plants can be quite impressive. The species is native to Asia and islands in the southwestern Pacific Ocean. Dog tail is not listed as an aggressive species although at least one of its relatives that I am familiar with prospers in disturbed areas. *Buddleia variabilis* was a popular flowering shrub at one time in and around Vancouver, British Columbia, where it is often mistakenly called a lilac. It is also known as the butterfly plant because these insects find the flowers particularly attractive. A few years ago a paved road was constructed in the local mountains to provide access to a ski development. Every year since, *Buddleia* has advanced along the roadside to a point where it is now several kilometers farther than when construction was begun. How far this advance will go is unknown; it may depend upon how hardy this species is at higher elevation where the chance of hard frost increases. Whether this species and the Hawaiian species represent actual pests or whether they are simply benign opportunists remains to be seen.

Emilia coccinea

This little plant is not likely to pose a major threat in the Hawaiian scene but it, and its relatives, are so widespread on the island, and their color so pronounced that it seemed reasonable to include at least one of them here. The genus *Emilia*, a member of the sunflower family, consists of about 100 species three of which are now widespread

Plate 184. *Emilia coccinea*, one of the "paint brush" plants.

weeds in the tropics (Mabberley, 1997). *Emilia coccinea* is illustrated in Plate (184). This species is also known by the common name "Flora's paintbrush." The other two species on the Hawaiian Islands, *E. fosbergii*, with brick red heads, and *E. sonchifolia* with lavender to pale purple flower heads, are also attractive plants. Needless to say, they have all been cultivated as garden ornamentals.

Fuchsia boliviana

Growing in the vicinity of a nasturtium patch in the Kokeʻe area is an attractive red fuchsia, *Fuchsia boliviana*. According to the *Manual*, this species of fuchsia, which is native to northwestern South America, was first collected in Hawaii in 1960. It appears to be naturalized in the Kokeʻe area where the photograph for Plate (185) was taken. Although this species doesn't seem to have become a pest on the Hawaiian Islands, at least two of its relatives have proved to be, one on St. Helena and one on

Réunion. On St. Helena, *Fuchsia coccinea* has been credited with the near extinction of the "large bellflower," *Wahlenbergia linifolia*, a member of Campanulaceae native to that island (*Plant Talk*, Jan. 2002). Also noted in that article was the current practice of removing the fuchsia from tree ferns as part of a program to control invasive species. *Fuchsia magellanica* is one of the more aggressive weeds targeted for removal on Réunion Island.

Plate 185. *Fuchsia boliviana,* with flowers up to 70 mm long (ca. 3 inches).

Hypochaeris—hairy cat's ear

This plant is not particularly aggressive; it might not even be noticed, but it does manage to get around. It will likely not present any major challenge to the island scene and it is probably not a threat to bring Hawaiian agriculture to its knees. It might even be listed among the alien species least likely to be a serious pest, although the effect of its presence on at least one endemic species might be worrisome. So why include it at all, especially following an introduction that used such words as loathsome and frightening? I include it because it is a plant that I know well; I have done battle with it for years in its efforts to claim my lawn as its territory. *Hypochaeris*, the hairy cat's ear, is a member of the sunflower family that has become widely distributed. (Its seeds are carried by a parachute like structure similar to the common dandelion.) It is obviously capable of prospering in a variety of habitats; it grows in my lawn at sea level, and appears to grow perfectly well at 10,000' among lava rocks on Haleakalā on East Maui. Plate (1) shows hairy cat's ear sharing a niche with *Tetramolopium humile*, one of the native species of *Tetramolopium* on the Hawaiian Islands. One wonders whether *T. humile* is robust enough to withstand competition from my lawn weed. Does hairy cat's ear have the staying power? Only time will tell. It should not escape one's notice that the occurrence of these two species together mirrors the state of the Hawaiian flora, 50% native and 50% alien, give or take a few percentage points.

Nasturtium

The garden nasturtium, with its golden, helmet-shaped flower, is familiar to most North American gardeners who would see it commonly in hanging baskets (or occasionally in salads in fancy restaurants). The botanical name for this plant is *Tropaeolum majus* and it belongs to its own small family, Tropaeolaceae, one of whose close relatives is the caper. The family occurs from Mexico to Chilé, but several cultivars have been in the horticultural trade as garden plants for many years. According

to the *Manual, T. majus* was first recognized as a naturalized species on
Maui before 1871. That this seem-
ingly innocuous plant could be a
potential problem often comes as a
surprise to folks who first hear about
it. Given its freedom in the
Hawaiian Island's forgiving climate,
however, it can spread at an alarm-
ing rate. Plate (186) illustrates this
species. Some years ago there was an
active campaign to eradicate this

Plate 186. *Tropaeolum majus,* the garden nasturtium
gone wild.

plant from Kīpuka Puaulu on the Big Island where it was growing well and beginning
to spread. During my most recent visit to that site (May, 2001) I saw no trace of the
plant, so the removal program seems to have been successful.

Polygonum

Polygonum (Polygonaceae) is a genus
having between 40 and 200 species
depending upon one's taxonomic
bias. Some of the species of weedy
Polygonum, the knotweeds, are widespread pests. The *Manual* lists seven
species all of which are naturalized
on the Hawaiian Islands, although
one of them, *P. glabrum,* has been
considered by some botanists as possibly being indigenous. The species
pictured in Plate (187) is *P. capitatum* growing on the open cinder field
in the vicinity of the Devastation
Trail on Hawai'i. It can also be found
scattered along the western rim of
Kīlauea Iki as the trail begins its
descent into the crater. The species
is characterized by a dark chevron
marking on its leaves as one can see
in Plate (188). The species is native

Plate 187. *Polygonum capitatum* invading the cinder
field along Desolation Trail, Island of Hawai'i.

Plate 188. *Polygonum capitatum* plants.

to the Himalayas and western China, has been cultivated on several of the Hawaiian Islands, and is now well established on Hawai'i. I have also seen this *Polygonum* species growing as a weed in gardens in southern California.

Solandra maxima

Although the authors of the *Manual* do not recognize this species as naturalized in the Hawaiian Islands, they do mention it in the introduction to their section on Solanaceae, the potato family, to which it belongs. H. D. Pratt (1998) suggests, however, that it may be naturalized in the hills above Honolulu (his book features a picture of the flower on the cover). The species is a liana with

Plate 189. *Solandra maxima*.

very large pale yellow flowers with five distinct purple lines. I have seen this spectacular plant along the road to Koke'e Park on Kaua'i where the photograph for Plate (189) was taken.

The aliens had help

When we enter a nature preserve on the Hawaiian Islands today we are strongly urged to clean our boots, socks, clothing, and packs of any adhering seeds or plant parts that we might inadvertently bring into a sensitive area. Plate (190) shows some seeds catching a ride on a passing sock. The potential harm that a single alien seed can do should be painfully obvious from what we have just seen

Plate 190. Hitchhikers on a sock.

above. This level of ecological sensitivity is a comparatively recent concern. There was a time in Hawaiian history when considerations of ecological sensitivity were not paramount. There were also some far-reaching decisions made based upon very peculiar ideas. One of these ideas, propounded by Harold Lyon, head of the Department of Botany and Forestry of the Hawai'i Sugar Planter's Association, and a man of some influence, believed that native forests are "doomed." Native species were thought not to be able to grow fast enough or well enough on disturbed sites to justify replanting them. The solution was to introduce new, fast growing forest trees. During the first half of the 20th century or so over 1,000 alien tree species were planted, roughly a tenth of the total number of alien species introduced! According to the story, as related by Cuddihy and Stone (1990), Lyon's favorite candidate for introduction was *Ficus*, e.g., the banyan tree, some 35 species of which were widely scattered. If this scatter shot approach was not bad enough, it is even more astounding that some of the seed dropped had not been identified to species! Many of these did not take to their new home; many other alien species did, however, and with great enthusiasm thus adding to the bigger problem. At best, this program might simply have been the product of profound naiveté, at worst, a shockingly irresponsible act of ecocide. Unfortunately, the replanting of ecologically inappropriate species continues today in several sectors of the North American forest industry, but that is a story for another day.

There is no room here, obviously, to go into detail about all of the aliens planted as trial reforestation species. A small sample should provide an idea of the magnitude of this program as well as the impact that it has had on the natural setting of the islands. *Psidium cattleianum*, the strawberry guava; *Tibouchina urvilleana*, the glory bush; *Grivellea robusta*, the silk oak; and most disastrously of all, *Myrica faya*, the firetree, were mentioned above and need not be revisited. Among the others were many species of *Eucalyptus*, some, such as *E. robusta*, the so-called swamp mahogany, were planted because they were fast growing as well as being potentially useful commercially, while others, *E. globulus*, the blue gum for example, were planted simply because they are fast growing. *Melaleuca quinquenervia*, the paperbark tree—the genus is related to *Eucalyptus*—was also planted despite the knowledge that it does not yield commercially useful wood. The Boy Scout Trail passes by a small grove of paperbark trees near the northern edge of the boggy area. A particularly troublesome tree that has become naturalized in several places in the islands is *Schinus terebinthifolius*, the so-called Brazilian pepper. This tree is a member of Anacardiaceae, the family that also houses the sumacs, poison ivy and poison oak, and mango. Brazilian pepper works its black magic by blocking the light from native species by overtopping them. The use of this plant's green foliage and red berries to make holiday decorations has earned it the common name Christmas berry. Another invading species that also works by producing heavy shade is the rose apple, *Syzygium jambos*. This genus is also a member of Myrtaceae along with the eucalypts and paperbark. The history of this alien has been well documented; the authors of the *Manual* inform us that it was brought by members of the crew of H. M. S. *Blonde* in 1825 from Rio de Janeiro because of its edible fruit. There is an endemic species of *Syzygium* (*S. sandwicensis*) on the Hawaiian Islands but its relationship to the South American tree has apparently not been studied. I have seen members of this genus planted as ornamentals on hotel grounds on Kaua'i.

CHAPTER FOUR

There is No Future in Extinction

Extinction has become a much talked about phenomenon over the past few years both in the common press and in the scientific literature. For example, the November 1, 2002 issue of the *Vancouver Sun* reprinted an article from *The Daily Telegraph* (London) pointing out that up to 47% of the world's plants face extinction. Earlier estimates had put the number closer to 13%, but two workers, Nigel Pitman of Duke University, and Peter Jorgensen of the Missouri Botanical Garden, suggested that the number should be much higher because the lower number was based on estimates that did not include threats to plants growing in tropical latitudes. Several recent discussions of extinction at a level appropriate for the general reader have appeared, as well as others that are aimed more at a professional audience. Among the former is the very popular *The Song of the Dodo: Island Biogeography in an Age of Extinctions* by David Quammen (1996) in which the extinction of both plants and animals is discussed in terms of the overall ecological inter-connectedness of natural systems. After having taken us to the restaurant at the end of the Universe (and back) Douglas Adams, along with Mark Carwardine, take us next to exotic locales to visit several endangered animal species in *Last Chance to See...* (1990). In *The Miner's Canary*, Niles Eldredge (1992), from the American Museum of Natural History in New York, laments "the drastic decline of songbirds who leave their ever-diminishing winter homes in tropical forests to brave their ever more hazardous migration routes up to their increasingly sparse and degraded breeding grounds in the northern hemisphere." A major theme in this treatment, as in the *Song of the Dodo*, is the disappearance of suitable habitats without which neither plants nor animals have much, if any, chance for reproduction. An early assessment of extinction status among plants and animals of the Americas was edited by G. Prance and T. Elias of the New York Botanical Garden in 1977. It was entitled *Extinction is Forever*. In *Extinction: Bad Genes or Bad Luck?* David Raup (1991), of the University of Chicago, presents a theoretical yet very readable account of the problems associated with deciding whether extinction occurs because of some inherent flaw in the genetic makeup of an organism, or whether it happens to beasts (in the broad sense) who happen to be in the wrong place at the wrong time. As in most either-or situations, it's likely some combination of the two.

It is useful to dwell on the latter ideas for a moment or two. It is a well known phenomenon that certain cells have a self-destruct (or at least fail-to-divide) command built into their genomes, that results in an organism with a limited period of existence. Interest in programmed cell death from the human perspective is an intensively studied area of research. Topics include, but are not limited to, genes that direct the formation of a protein called telomerase that repairs the "frayed ends of chromosomes," to use the words of Matt Ridley in his 1999 book *Genome: The Autobiography of a Species in 23 Chapters*. (The species he talks about is *Homo sapiens*.) Ridley also discusses current interest in cancer cells that can be instructed to commit suicide, i.e., stop dividing.

It seems highly unlikely, however, that specific intrinsic factors could lead to extinction of a species as a whole. The factors mentioned above are of significance to individual organisms but there is likely sufficient variation in expressing these traits within a population (or species) that the larger assembly of organisms is not affected. That species die, however, is well known; the vast majority of species that ever existed on the Earth no longer exist. In fact, it has been estimated that the average life of a species is approximately two million years give or take a few millennia. Some, of course, persist for much longer periods of time while others may have only a fleeting existence. The one thing that is certain is that no species exists forever. What causes a species to become extinct, other than by the heavy foot of humankind, that is? Some possible scenarios were discussed by David Raup, one of which is of particular importance in understanding extinction in island systems. It all has to do with the level of genetic variation within the species—we have to focus on the group here—and how many individuals comprise the species. Small populations are particularly vulnerable since loss of a few individuals can reduce the number of individuals below the minimum number needed for breeding, i.e., animals finding an increasingly scarce mate, or plants producing sufficient pollen to saturate the local environment. Species consisting of small numbers of individuals are also at higher risk from changes in their natural environment such as change in rainfall pattern or average daily temperature. A species that consisted of a relatively few individuals would lack the buffering capacity of a larger group; that is, a larger group would be able to absorb loss of a fair number of individuals without being brought to the edge of extinction. In other words, death of 20% (say) of the members of a group could have a disastrous impact on a small species compared to a large one. Since species on islands tend to consist of fewer members than continental species, the problem of minimum reproductive number and sensitivity to environmental change is acute.

You may recall a number of cases from earlier chapters where we met species whose numbers have been reduced to a dozen or so individuals. Under threat, a few individuals may survive, but they will carry only a small fraction of genetic variation. A second bad year for whatever reason can easily push the species over the brink. Add to the natural threats, over zealous collection of specimens by earlier botanists and destruction of habitats by human activities, both historical and current, and one has a formula for disaster. It is only necessary to recall the case of extinct and endangered species of mints on the Hawaiian Islands in Chapter Two to see a vivid example of the problem. But steps can be taken, if not to eliminate the problem completely, at least to recognize it and try to devise a remedy. We can look at some of these activities next.

Over the last few decades there has been an increasing awareness of the problems facing the Earth's natural environment. Some scientists have expressed the opinion that we are witnessing a "species extinction crisis," and that if drastic conservation measures aren't taken the situation will worsen to the point of no return. The term "biodiversity" is seen in print and heard discussed more and more. How does one measure biodiversity? Is it simply the total number of all recognized species of plants, animals, and microorganisms on Earth, plus a guess at how many others there might be? What about richness of life forms per unit area? Might it be possible to identify some keystone species or group of species and monitor their growth over time? A factor that might be considered in assessing importance of a given group of plants is the level of genetic diversity exhibited within a defined area. Another possibility would be to assess the condition of various ecosystems and the severity of threats to them. Conservation International (CI), an international body of concerned scientists, decided that two factors should be taken into account in defining an ecoregion in need of conservation efforts: levels of vascular plant endemism, and degree of threat to the region. A "hotspot" was defined as an area that would have at least 0.5% of total global vascular plant diversity and at least 1,500 endemic species. Plants were selected as the key organisms because there is reliable information available and, they are, after all, the life forms upon which all other life of Earth depends. The second criterion for recognition of an area as a hot spot is the degree of threat: 25% or less of its original plant cover remains intact. As of the April, 2000 issue of *Plant Talk* (Villa-Lobos, 2000), 25 hot spots had been identified. These areas represent 1.44% of the total land area of the Earth but 44% of all vascular plant species and 35% of the four vertebrate groups (birds, mammals, reptiles, and amphibians). Two tables from the *Plant Talk* article that list the hot spots and their corresponding levels of vascular plant endemism have been combined in Table (1) below. It seems entirely unnecessary to emphasize how much of the Earth's biological wealth has

Table 1. Hot spots. Numbers of species of vascular plants and level of endemism (modified from Villa-Lobos, 2000).

Hotspot	Species	Endemics	% Endemism
Tropical Andes	45,000	20,000	44.4
Mesoamerica	24,000	5,000	20.8
Caribbean	12,000	7,000	58.3
Atlantic Forests of Brazil	20,000	8,000	40.0
Chocó/ Darién/ Western Ecuador	9,000	2,250	25.0
Brazilian Cerrado	10,000	4,400	44.0
Central Chilé	3,429	1,605	46.8
California Floristic Province	4,426	2,125	48.0
Madagascar (and Indian Ocean Islands)	12,000	9,704	80.9
Eastern Arc Mtns. & Coastal Forests of Kenya & Tanzania	4,000	1,500	35.0
Western African Forests	9,000	2,250	25.0
Cape Floristic Province	8,200	5,682	69.3
Succulent Karoo (African Cape area)	4,849	1,940	40.0
Mediterranean Basin	25,000	13,000	52.0
Caucasus	6,300	1,600	25.4
Sundaland	25,000	15,000	60.0
Wallacea	10,000	1,500	15.0
Philippines	7,620	5,832	76.5
Indo-Burma	13,500	7,000	51.9
South-Central China	12,000	3,500	29.2
Western Ghats/Sri Lanka	4,780	2,100	45.6
Southwestern Australia	5,469	4,331	79.2
New Caledonia	3,332	2,551	76.8
New Zealand	2,300	1,865	81.1
Polynesia/Micronesia	6,557	3,334	50.8

been relegated to so small a total area! Is anything being done to protect what little there seems to be left? The answer is yes, but is it enough? The author of the article points out that 38% of the hotspots are protected in some way, either through parks or reserves. Although this suggests that the organisms in these protected areas are safe, the reality is often a different matter. Safeguards against changes in rules governing these areas are essential to protect them from future attempts to allow development or resource extraction to occur. Note that the Hawaiian Islands are included under Polynesia/Micronesia in this compilation. The entire subject is far too vast to tackle here so we will return to the Hawaiian Islands to see some of the approaches being taken to confront the local issues.

A major aim of international programs dealing with biological diversity and the hotspots is to set aside large areas as parks or reserves. Visitors to the Hawaiian Islands are privileged in having the systems of national parks where large tracts of land have been set aside in perpetuity: Hawaiʻi Volcanoes National Parks on Hawaiʻi (Kīlauea) and Maui (Haleakalā), the Kalaupapa National Historical Park on the north coast of Molokaʻi. Many State parks have also been established such as the Nā Pali Coast State Park on the north coast of Kauaʻi, Polihale State Park on the west coast of Kauaʻi, and Mākua-Kaʻena State Park on the northwestern tip of Oʻahu, among others. Natural reserves have also been established such as the Hanawi Natural Area Reserve on East Maui and the Scientific Research Reserve in the Kīpahulu Valley, East Maui. Paramount among the gardens, however, is the National Tropical Botanical Garden which has facilities at several places on the islands: the Allerton or Lāwaʻi Garden located on Kauaʻi's south coast, the Limahuli Garden on the north coast of Kauaʻi, and Kahanu Garden at Hana on East Maui. The magnificently situated Limahuli Garden includes several species otherwise difficult to find. In Honolulu a visitor would also be well rewarded by paying a visit to the Lyon Arboretum, which lies in the Upper Manoa Valley.

Recognition of a potential extinction problem depends on continuing field observations coupled with an extensive knowledge of what is known about the distribution of a given group of plants and how their range may have changed over time. If a given species is considered at risk it is possible, using current techniques for determining levels of genetic variation, to allow scientists to assess the degree of genetic risk and to plan a conservation strategy. Seeds or other plant parts can be collected for cultivation with plans for eventual reintroduction. An example of mass cultivation was shown in Chapter Two with a photograph of a crop of *Brighamia insignis* (Plate 48) being maintained at the Kīlauea Lighthouse site on Kauaʻi. Further efforts to keep it and its related species from harm have involved artificial pollination in the field, often requiring extraordinary efforts to reach plants growing on cliff faces. In some situations seeds do not germinate well in the glass house or seedlings may not be robust. In these extreme cases techniques involving culturing of plant embryos or other tissues can be used to provide new individuals for further cultivation or out-planting. The use of tissue culturing may sound a little like a science fiction story but it is an approach widely used for plant propagation where seeds don't cooperate or where a high level of uniformity is desired, or in more extreme cases, where some modified sequence of DNA is being inserted into an organism using genetic engineering techniques.

Another major threat to island species is the presence of invasive weeds, several of which were featured in Chapter Three. Hands-on weeding is labor intensive but it does work. Groups of volunteers work along side park personnel to locate and destroy individual plants. For example, there are efforts to remove fire trees (*Myrica faya*) in the Devastation Trail area near **Kīlauea** Iki. Many of the invasive Melastomataceae, e.g., Koster's curse (*Clidemia hirta*) and green cancer (*Miconia calvescens*), can be dealt with in this way as well. Hikers are encouraged to be on the lookout for these nasty weeds and to dig them out themselves if possible or report their position to local authorities (see Chapter Three for contact numbers). Considerable attention is also paid to plants that people put in their gardens and around their homes for decorative purposes. Just how dangerous that can be was illustrated through the introduction of the beautiful, large leafed *Miconia calvescens* which soon became one of the most invasive weeds on Hawai'i.

Most readers could be forgiven for thinking that ferns are unlikely candidates for listing as invasive weeds. In an article titled "Alien Ferns in Hawai'i" K. A. Wilson (1996) reports that over 260 species of alien ferns and fern allies are in cultivation in botanical gardens and arboretums in the islands and that they pose a real threat to native ferns. An aggressive species of fern can be as invasive as any flowering plant; one only has to see the extent that bracken fern can cover an area following fire or other disturbance (Plate 117).

Herbicide sprays can be used but the obvious problem with this approach is the possible sensitivity of native plants to the chemicals. An approach that has been tried worldwide and to some extent in the Hawaiian Islands involves introduction of an insect or pathogen from the weed's native area in the hope that it will continue its good work. As related in Chapter Three, this approach can backfire if the imported insect takes a liking to something else, or does not itself grow well and reproduce in its new home.

What can the occasional visitor do to help preserve the natural riches of the islands? Other than joining a group and spend part of a vacation as a volunteer weeder, the most obvious answer is not a great deal, but that answer refers only to the hands-on part of the battle. The most important thing to do is to become informed. Take an interest in the native plants and animals, visit botanical gardens, unspoiled natural areas, and arboretums. Explore some of the forest trails on the islands—in other words, get wet and muddy—and make an effort to appreciate what you are seeing. Remember that what you are looking at may occur nowhere else, and unless we all pitch in to help save it, it may end up occurring nowhere at all.

All taxonomic and evolutionary studies of plants start with morphology, the detailed examination of plant structure. By comparing suites of characters of a group of plants, skilled botanists can sort them into groups according to similarities. Experience then allows the workers to postulate relationships with other known species and to speculate on whether the suite of characters reflects an evolutionary primitive or advanced position. (Some contemporary botanists take those two terms as pejorative and prefer generalized vs. specialized features.) Important information can also be obtained by looking at the interior architecture of a plant—the field of plant anatomy—including such features as its vascular system, the way its photosynthetic apparatus is arranged, and such specialized features as oil cells, or crystalline inclusions.

A plant's chromosomes can also provide important clues to relationships. The number of chromosomes can be very useful as can their shapes and sizes. The term "chromosome number" refers to the number of pairs of chromosomes present in vegetative (non-reproductive) cells; it is routinely determined by looking for chromosomes in the nuclei of rapidly growing cells such as root tips. This value is referred to as the "diploid" number and is written as $2n = 12$, for example. (Chromosome number can also be determined by looking at reproductive cells after the chromosome pairs have split in the formation of gametes; this is referred to as the "haploid" number and would be written in this case as $n = 6$.) Deviations from the diploid chromosome number of a species can serve as a clear indication that something of evolutionary significance may have happened. For example, chromosome numbers may change incrementally in closely related species resulting in a series, again using $2n = 12$ as an example, consisting of species with $2n = 10$. Considering that $2n = 12$ is thought to be the fundamental chromosome number for this particular genus, one would conclude that the lower number represents a derived species and therefore of more recent origin. Entire sets of chromosomes can double, in the case of our example one might find a species with 24 chromosomes ($2n = 24$). A species with this number would be called a "tetraploid." A species within our hypothetical genus with this number of chromosomes would also be interpreted as being derived. Chromosome number, therefore, can provide an overview of the direction of evolutionary change within a group of related species (or genera). Recall that diploid and tetraploid species of *Scaevola* exist on the Hawaiian Islands and that this had a bearing on hypotheses of their relationships.

Members of a species can, in theory at least, exchange genes with any other member of the species in a process that we know as "crossing" or "outcrossing." Open exchange of genes within a species assures an equitable distribution of genetic resources within the group and limits the amount of "inbreeding" or "selfing."

Inbreeding can lead to the accumulation of deleterious genes which can result in serious problems with growth and other processes. Thus, understanding the reproductive biology of a species under scrutiny may provide useful insights into its evolutionary history.

Chromosomes are subject to more or less regular changes in structure that result in altered DNA sequences. The simplest involves a change in the structure of a particular nucleotide base, one of the links in the genetic chain, resulting in what is called a point mutation. Most point mutations have little or no overall effect on the plants in which they occur. Although of no apparent harm (or use) to the plants in which they occur, these changes nonetheless have the potential of providing scientists with markers that can be detected through the use of certain specialized techniques described below. Other changes can involve the interchange of pieces of DNA between different chromosomes, or they may involve inversion of a segment of DNA within a single chromosome. These changes do not involve loss of DNA but they do change the order of appearance of the building blocks (nucleotide base) within one or more chromosomes. Whereas the change in a DNA sequence caused by structural rearrangements of chromosomes requires biochemical analysis, the structural changes themselves are visible through the microscope using a variety of techniques.

The next thing of interest to us is what happens when one population of a species becomes isolated from all of the others. The isolated population will continue to accumulate mutations but, owing to its isolation, will have significantly reduced (or no) opportunities to exchange them with other populations. The obvious result of this process is that the isolated population will begin to develop a unique suite of features. In time, the accumulation of differences between this population and all others will result in sufficient morphological divergence that it may no longer be recognized as a member of the original species. In addition to changes in physical structure, other processes can be affected by divergence, among which is the capacity to interbreed with members of the original species. A botanist interested in the degree of divergence between a parent species and its daughter species may study the capacity of the two to interbreed. The rule of thumb in this is simple: the greater the separation in evolutionary time (number of generations) the less likely the two species will recognize each others' genetic makeup. The reader may recall that breeding studies of this sort played a major role in working out relationships between members of the silversword alliance and their putative mainland ancestors. It is in studies of this sort, where crosses between distant relatives are examined, that the structural changes in chromosomes mentioned above become a potentially useful set of markers. The reproductive lives of potential colonizers, whether they are capable of self pollination or whether they are obligate outcrossers, dictate to a significant degree their success in the new island setting.

Mentioned above was the idea that genes accumulate mutations over time. Individuals that experience mutations in certain key spots in their genes die, whereas most mutations result in little more than a minor change in the structure of particular proteins. Structural changes in proteins often result in changes in their electrical charge, which provides a means of comparing protein profiles from different individuals. The technique is called protein electrophoresis and involves separating a mixture of proteins in an electric field (the distance migrated on a neutral medium by a protein depends upon its net electrical charge). The resulting banding patterns can be compared and provide a quantitative measure of genetic variation within a population or species, between two populations or species, and among several populations or species. An invaluable contribution to our understanding of evolutionary direction, using electrophoretic data, is the fact that daughter species will exhibit only a fraction of the genetic variation present in their parent species. In other words, a seed being carried to another site or another island carries a unique array of gene mutations (and hence experimentally observable protein profiles), and since it may be the only seed to make the trip, the only genetic information available in the new site is that which was packed inside that seed. This is akin to saying that you would have only those clothes to wear on your island holiday that you took with; those left home in the closet are no longer relevant.

Bursting onto the scene a few years ago were techniques that allowed isolation of DNA using very small amounts of plant material (100 milligrams or so), and the means by which the sequence of building blocks (nucleotide bases) in individual genes could be determined. It became possible, with comparative ease, to determine exactly where each of the accumulated mutations was located in the DNA strand. Examination of the sequences of a particular gene from two putatively related species allows one to estimate how long it has been since the two species shared a common ancestor. It is also known that different genes tend to accumulate mutations at different rates. If the times of divergence of two—or ideally more—independently evolving genes are in agreement, it is possible to make a much stronger case for a particular relationship.

BIBLIOGRAPHY

Abbott, I. A. 1992. Lā'au Hawai'i. Traditional Hawaiian Uses of Plants. Bishop Museum Press, Honolulu, Hawai'i.

Anonymous. Information brochure for Kipuka Huluhulu. State of Hawai'i Dept. of Land and Natural Resources, Division of Forestry and Wildlife. (Available at the site.)

Anonymous. *Limahuli Garden.* A window to ancient Hawai'i. National Tropical Botanical Garden, Hā'ena, Kaua'i. (Available at the site.)

Adams, D. and Carwardine, M. 1990. Last Chance to See... . Stockard Publishing Co., Toronto.

Baldwin, B. G., and Wessa, B. L. 2000. Origin and relationships of the tarweed—silver sword lineage (Compositae—Madiinae). American Journal of Botany 87: 1890-1908.

Ballard, Jr., H. E., and Sytsma, K. J. 2000. Evolution and biogeography of the woody Hawaiian violets (*Viola*, Violaceae): arctic origins, herbaceous ancestry and bird dispersal. Evolution 54: 1521-1532.

Bohm, B. A. 1998. Secondary compounds and evolutionary relationships of island plants. *In* T. E. Stuessy and M. Ono (eds.) Evolution and Speciation of Island Plants. Cambridge University Press, Cambridge, UK.

Bohm, B. A., and Koupai-Abyazani, M. R. 1994. Flavonoids and condensed tannins from leaves of *Vaccinium reticulatum* and V. *calycinum* (Ericaceae). Pacific Science 48: 458-463.

Bowen, E., and Van Vuren, D. 1997. Insular endemic plants lack defenses against herbivores. Conservation Biology 11: 1249-1254.

Carlquist, S. 1959. Studies on Madinae: anatomy, cytology, and evolutionary relationships. Aliso 4: 171-236.

Carlquist, S. 1967. The biota of long distance dispersal. V. Plant dispersal to Pacific Islands. Bulletin of the Torrey Botanical Club 94: 129-162.

Carlquist, S. 1974. Island Biology. Columbia University Press, New York.

Carlquist. S. 1980. Hawaii, a natural history. Geology, climate, native flora and fauna above the shoreline. 2nd edn. Pacific Tropical Botanical Garden, Lawai, Hawaii.

Carlquist, S., Baldwin, B., and Carr, G. (eds.) (2003) Tarweeds and Silverswords. Evolution of the Madiinae (Asteraceae). Missouri Botanical Garden Press, St.Louis.

Carr, G. D. 1985. Monograph of the Hawaiian Madiinae (Asteraceae): *Argyroxiphium*, *Dubautia*, and *Wilkesia*. Allertonia 4: 1-123.

Carr, G. D., Baldwin, B. G., and Kyhos, D. W. 1996. Cytogenetic implications of artificial hybrids between the Hawaiian silversword alliance and North American tarweeds (Asteraceae:Heliantheae—Madiinae). American Journal of Botany 83: 653-660.

Carr, G. D., and Kyhos, D. W. 1981. Adaptive radiation in the Hawaiian silversword alliance (Compositae—Madiinae). I. Cytogenetics of spontaneous hybrids. Evolution 35: 543-556.

Carr, G. D., and Kyhos, D. W. 1986. Adaptive radiation in the Hawaiian silversword alliance (Compositae—Madiinae). II. Cytogenetics of artificial and natural hybrids. Evolution 40: 959-976.

Carson, H. L. and Clague, D. A. 1995. Geology and Biogeography of the Hawaiian Islands. *In* Hawaiian Biogeography: Evolution on a Hot Spot Archipelago. W. L. Wagner and V. A. Funk (eds.) Smithsonian Institution Press, Washington, D.C.

Costello, A. and Motley, T. J. 2001. Molecular systematics of *Tetramolopium*, *Munroidendron* and *Reynoldsia sandwicensis* (Araliaceae) and the evolution of superior ovaries in *Tetraplasandra*. Edinburgh Journal of Botany 58: 229-242.

Cronk, Q. C. B., and Fuller, J. L. 2001. Plant invaders. The threat to natural ecosystems. Earthscan Publications Inc., London, UK.

Cruickshank, D. P. 1986. Mauna Kea. A Guide to the Upper Slopes and Observatories. University of Hawaii, Institute for Astronomy, Honolulu.

Cuddihy, L. W., and Stone, C. P. 1990. Alteration of Native Hawaiian Vegetation. Effects of Humans, Their Activities and Introductions. Cooperative National Park Resources Study Unit, University of Hawaii. Distributed by University of Hawai'i Press, Honolulu.

Culliney, J. L. and Koebele, B. P. 1999. A Native Garden. How to Grow and Care for Island Plants. University of Hawai'i Press, Honolulu, Hawai'i.

Daws, G. 1968. Shoals of Time. A History of the Hawaiian Islands. University of Hawai'i Press, Honolulu.

DeJoode, D. R. and Wendel, J. F. 1992 Genetic diversity and origin of the Hawaiian Islands cotton, *Gossypium tomentosum*. American Journal of Botany 79: 1311-1319.

Devine, R. 1998. Alien Invasion. America's Battle with Non-native Animals and Plants. National Geographic Society, Washington, D.C.

Eldredge, N. 1992. The Miner's Canary: Unraveling the Mysteries of Extinction. Virgin Books, London, UK.

Fosberg, F. R. 1948. Immigrant plants in the Hawaiian Islands. II. Occasional Papers. University of Hawaii 46: 1-17.

Fosberg, F. R. and Herbst, D. 1975. Rare and Endangered Species of Hawaiian Vascular Plants. Allertonia 1: 1-72.

Ganders, F.R., Berbee, M., and Pirseyedi, M. 2000. ITS Base sequence phylogeny in *Bidens* (Asteraceae): evidence for the continental relatives of Hawaiian and Marquesan *Bidens*. Systematic Botany 25: 122-133.

Gemmill, C. E. C., Allan, G. J., Wagner, W. L., and Zimmer, E. A. 2002. Evolution of insular Pacific *Pittosporum* (Pittosporaceae): origin of the Hawaiian radiation. Molecular Phylogenetics and Evolution 22: 31-42.

Gemmill, C. E. C., Ranker, T. A., Ragone, D., Perlman, S. P. and Wood, K. R. 1998. Conservation genetics of the endangered endemic Hawaiian genus *Brighamia* (Campanulaceae) American Journal of Botany 85: 528-539.

Givnish, T. J., Knox, E., Patterson, T. B., Hapeman, J. R., Palmer, J. D., and Sytsma, K. J. 1996a. The Hawaiian lobelioids are monophyletic and underwent a rapid initial radiation roughly 15 million years ago. American Journal of Botany 83:159 (Supplement).

Givnish, T. J., Sytsma, K. J., Patterson, T. B., and Hapeman, J. R. 1996b. Comparison of patterns of geographic speciation and adaptive radiation in *Cyanea* and *Clermontia* (Campanulaceae) based on cladistic analysis of DNA sequence and restriction-site data. American Journal of Botany 83:159 (Supplement).

Givnish, T. J., Sytsma, K. J., Smith, J. F., and Hahn, W. J. 1995. Molecular evolution, adaptive radiation, and geographic speciation in *Cyanea* (Campanulaceae, Lobelioideae). *In* Hawaiian Biogeography: Evolution on a Hot Spot Archipelago. W. L. Wagner and V. A. Funk (eds.) Smithsonian Institution Press, Washington, D.C., pp. 288-337.

Gutmanis, J. 2001. Hawaiian Herbal Medicine. Kāhuna Lāʻau Lapaʻau. Island Heritage Publishing, Waipahu, Hawaiʻi. Handy, E. S. C., Handy, E. G., and Pukui, M. K. 1991. Native Planters in Old Hawaii. Their Life, Lore, and Environment. Bishop Museum Press, Honolulu, Hawaii.

Hillebrand, W. 1888. Flora of the Hawaiian Islands: a description of their phanerogams and vascular cryptogams. Carl Winter, Heidelberg; Williams & Norgate, London; B. Westermann & Co., New York. (Facsimile ed., 1965, Hafner Publ. Co., New York; Facsimile ed., 1981, Lubrecht & Cramer, Monticello, New York.)

Hobdy, R. 1993. Lānaʻi—A case study: The Loss of Biodiversity on a Small Hawaiian Island. Pacific Science 47: 201-210.

Howarth, D. G., Gustafsson, M. H. G., Baum, D. A., and Motley, T. 2003. Phylogenetics of the genus *Scaevola* (Goodeniaceae): implications for dispersal patterns across the Pacific Basin and colonization of the Hawaiian Islands. American Journal of Botany 90: 915-923.

Juvik, S. P., and Juvik, J. O. (eds.) 1998. Atlas of Hawaii. 3rd edn. University of Hawaiʻi Press, Honolulu.

Keck, D. D. 1936. The Hawaiian silverswords: systematics, affinities, and phytogeographic problems of the genus *Argyroxiphium*. Occasional Papers of the Bernice P. Bishop Museum 11: 1-38.

Kepler, A. K. 1988. Haleakalā: A Guide to the Mountain. Mutual Publishing. Honolulu, Hawaiʻi.

Kepler, A. K. 1990. Trees of Hawaiʻi. University of Hawaiʻi Press, Honolulu, Hawaiʻi.

Kepler, A. K. 1991. Majestic Molokaʻi. Mutual Publishing. Honolulu, Hawaiʻi.

Kepler, A. K. 1995. Maui's Floral Splendor. Mutual Publishing. Honolulu, Hawaiʻi.

Kepler, A. K. 1997. Hawai'i's Floral Splendor. Mutual Publishing. Honolulu, Hawai'i.

Kepler, A. K. 1998. Hawaiian Heritage Plants. University of Hawai'i Press, Honolulu, Hawai'i.

Koske, R. E., and Gemma, J. N. 1990. VA Mycorrhizae in strand vegetation of Hawaii: evidence for long-distance codispersal of plants and fungi. American Journal of Botany 77: 466-474.

Krauss, B. H. 1993. Plants in Hawaiian Culture. University of Hawai'i Press, Honolulu.

Lamb, S. H. 1981. Native Trees and Shrubs of the Hawaiian Islands. Sunstone Press, New Mexico.

Lamoureux, C. H. 1976. Trailside Plants of Hawai'i's National Parks. Hawaiian Natural History Association and National Park Service

Lebot, V., and Lévesque, J. 1989. The origin and distribution of kava (*Piper methysticum* Forst. f., Piperaceae): a phytochemical approach. Allertonia 5: 223-281.

Lesko, G. L. and Walker, R. B. 1969. Effect of seawater on seed germination of two Pacific atoll beach species Ecology 50: 730-734.

Liittschwager, D., and Middleton, S. 2001. Remains of a Rainbow. Rare Plants and Animals of Hawaii. National Geographic Society, Washington, D.C.

Lindqvist, C., and Albert, V. A. 2002. Origin of the Hawaiian endemic mints within North American *Stachys* (Lamiaceae). American Journal of Botany 89: 1709-1724.

Lowrey, T. K. 1986. A biosystematic revision of Hawaiian *Tetramolopium* (Compositae:Asterae). Allertonia 4: 203-265.

Lowrey, T. K. 1995. Phylogeny, adaptive radiation, and biogeography of Hawaiian *Tetramolopium*. *In* Hawaiian Biogeography. Evolution on a Hot Spot Archipelago. W. L. Wagner and V. A. Funk (eds.) Smithsonian Institution Press, Washington, D.C., pp. 195-220.

Mabberley, D. J. 1997. The Plant-Book, 2nd edn., Cambridge University Press, Cambridge, UK.

Macdonald, G. A., Abbott, A. T., and Peterson, F. L. 1990. Volcanoes in the Sea. The Geology of Hawaii, 2nd edn. University of Hawai'i Press, Honolulu.

Macdonald, G. A., and Hubbard, D. H. 2001. Volcanoes of the National Parks in Hawai'i. Hawai'i Natural History Association and National Park Service, Hawaii National Park, Hawai'i. [2001 Update prepared by C. Heliker and D. Swanson.]

MacLeod, R. and Rehbock, P. E. (eds.) 1994. Darwin's Laboratory. Evolutionary Theory and Natural History in the Pacific. University of Hawai'i Press, Honolulu.

Macrae, J. 1825. With Lord Byron at the Sandwich Islands in 1825. Being Extracts from the MS Diary of James Macrae, Scottish Botanist. Reprinted in 1972 by The Petrograph Press, Hilo, Hawaii, with a Foreword by Wm. F. Wilson (orig. 1922).

Marchant, Y., Turjman, M., Flynn, T., Balza, F., Mitchell, J. C., and Towers, G. H. N. 1985. Identification of psoralen, 8-methoxypsoralen, isopimpinellin, and 5,7-

dimethoxycoumarin in *Pelea anisata* H. Mann. Contact Dermatitis 12: 196-199.

Marr, K. L. and Bohm, B. A. 1997. A taxonomic revision of the endemic Hawaiian *Lysimachia* (Primulaceae) including three new species. Pacific Science 51: 254-287.

Meyen, F. J. F. 1831. A botanist's visit to Oahu in 1831. Being the journal of Dr. F. J. F. Meyen's travels on observations about the island of Oahu. Translated by Astrid Jackson, 1981, Press Pacifica, Ltd., Kailua, Hawaii.

Meyrat, A. K., Carr, G. D., and Smith, C. W. 1984. A morphometric analysis and taxonomic appraisal of the Hawaiian silversword *Argyroxiphium sandwicense* DC. (Asteraceae). Pacific Science 37: 211-225.

Mummenhoff, K., Brüggemann, H., and Bowman, J. L. 2001. Chloroplast DNA phylogeny and biogeography of *Lepidium* (Brassicaceae). American Journal of Botany 88:2051-2063.

Neal, M.C. 1965. In Gardens of Hawai'i. Bishop Museum Special Publication 50. Bishop Museum Press, Honolulu, Hawai'i.

Palmer, D. D. 2003. Hawai'i's Ferns and Fern Allies. University of Hawai'i Press, Honolulu.

Pax, D. L., Price, R. A., and Michaels, H. J. 1997. Phylogenetic position of the Hawaiian geraniums based on *rbcL* sequences. American Journal of Botany 84:72-78.

Powell, E. A., and Kron, K. A. 2002. Hawaiian blueberries and their relatives—a phylogenetic analysis of *Vaccinium* sections *Macropelma*, *Myrtillus*, and *Hemimyrtillus* (Ericaceae). Systematic Botany 27: 768-779.

Prance, G. T. and Elias, T. S. (eds.) 1977. Extinction is Forever. The New York Botanical Garden, Bronx, New York.

Pratt, H. D. 1998. A Pocket Guide to Hawai'i's Trees and Shrubs. Mutual Publishing, Honolulu.

Price, J. P., and Clague, D. A. 2002. How old is the Hawaiian biota? Geology and phylogeny suggest recent divergence. Proceedings of the Royal Society of London, Ser. B 269: 2429-2435.

Pukui, M. K. and Elbert, S. H. 1986. Hawaiian Dictionary. University of Hawai'i Press, Honolulu.

Purugganan, M. D., Friar, E. A., and Robichaux, R. H. 2002. Hawaiian Silversword Alliance: Evolution and Conservation. Abstract of talk at the 49th Annual Systematics Symposium, Missouri Botanical Garden, St. Louis.

Robichaux, R., Bakutis, A., Bergfeld, S., Bio, K., Bruegmann, M., Canfield, J., Friar, E., Jacobi, J., Moriyasu;, P., Perry, L., Rubenstein, T., Tunison, T., and Warshauer, F. 2001. Reintroduction of the Endangered Mauna Loa Silversword, *Argyroxiphium kauense* (Asteraceae). Abstract from talk at: Society for Conservation Biology Meetings, 2001, Hilo, Hawai'i.

Quammen, D. 1996. The Song of the Dodo. Island Biogeography in an Age of Extinction. Scribner, New York.

Raup, D. M. 1991. Extinction. Bad Genes or Bad Luck? W. W. Norton & Co., New York.

Rauzon, M. J. 2002. Isles of Refuge. Wildlife and History of the Northwestern Hawaiian Islands. University of Hawai'i Press, Honolulu.

Ridley, M. 1999. Genome: The Autobiography of a Species in 23 Chapters. Harper Collins, New York.

Scheuer, P. J. 1955. The constituents of mokihana (*Pelea anisata*). Chemistry & Industry 1257-1258.

Scheuer, P. J., and Hudgin, W. R. 1964. Major constituents of the essential oil of *Pelea christophersenii*. Perfumery and Essential Oil Record 55: 723-724.

Skottsberg, C. 1972. The genus *Wikstroemia* in the Hawaiian Islands. Cited by Wagner et al. (1999) in the Manual.

Smith, R. 1993. Hiking Maui. 6th edn. Hawaiian Outdoor Adventures Publications, Huntington Beach, California.

Sohmer, S. H. and Gustafson, R. 1987. Plants and Flowers of Hawai'i. University of Hawai'i Press, Honolulu.

Staples, G. W., and Cowie, R. H. 2001. Hawai'i's Invasive Species. Mutual Publishing and the Bishop Museum Press, Honolulu.

Stoddart, D. R. 1994. "This coral episode." Darwin, Dana, and the coral reefs of the Pacific. *In* Darwin's Laboratory. Evolutionary Theory and Natural History in the Pacific. R. MacLeod and P. E. Rehbock (eds.) University of Hawai'i Press, Honolulu.

Stone, C. P. and Stone, D. B. 1989. Conservation Biology in Hawai'i. University of Hawai'i Press, Honolulu.

Thornton, I. 1996. Krakatau. The Destruction and Reassembly of an Island Ecosystem. Harvard University Press, Cambridge, Massachusetts.

Valier, K. 1995. Ferns of Hawai'i. University of Hawai'i Press, Honolulu, Hawai'i.

Van der Pijl L. 1972. Principles of dispersal in higher plants. Springer-Verlag, New York.

Vargas, P., Baldwin, B. G., and Constance, L. 1998. Nuclear ribosomal DNA evidence for a western North American origin of Hawaiian and South American species of *Sanicula* (Apiaceae). Proceedings of the National Academy of Science, U.S.A. 95:235-240.

Vargas, P., Baldwin, B. G., and Constance, L. 1999. A phylogenetic study of *Sanicula* sect. *Sanicoria* and S. sect. *Sandwicensis* (Apiaceae) based on nuclear rDNA and morphological data. Systematic Botany 24:228-248.

Villa-Lobos, J. 2000.Biodiversity Hotspots for Conservation Planning. Plant Talk No. 21: 30-32

Wagner, W. H., Jr. 1995. Evolution of Hawaiian Ferns and Fern Allies in Relation to the Conservation Status. Pacific Science 49: 31-41.

Wagner, W. L. 1996. *Scaevola hobdyi* (Goodeniaceae), an enigmatic new species from West Maui. Novon 6: 225-227.

Wagner, W. L., Herbst, D. R., and Sohmer, S. H. 1990, 1999. Manual of the Flowering Plants of Hawai'i. University of Hawai'i Press, Honolulu.

Wagner, W. L. and Funk, V. L. (eds.) 1995. Hawaiian Biogeography. Evolution on a Hot Spot Archipelago. Smithsonian Institution Press, Washington, D.C.

Wagner-Wright, S. (ed.) 1999. Ships, Furs, and Sandalwood. A Yankee Trader in Hawai'i, 1823-1825. University of Hawai'i Press, Honolulu.

Westerbergh, A., and Saura, A. 1994. Genetic differentiation in endemic *Silene* (Caryophyllaceae) on the Hawaiian Islands. American Journal of Botany 81: 1487-1493.

Whistler, W. A. 1980. Coastal Flowers of the Tropical Pacific. Oriental Publishing Co.

Wiesel, D., and Stapleton, F. 1992. Aloha O Kalapana. Bishop Museum Press, Honolulu.

Wilson, K. A. 1996. Alien ferns in Hawai'i. Pacific Science 50: 127-141.

Wilson, J. T. 1963. A possible origin of the Hawaiian Islands. Canadian Journal of Physics 41: 863-870.

Winchester, S. 2003. Krakatoa. Harper Collins Publishers. New York.

Wright, S. D., Yong, C. G., Dawson, J. W., Whittaker, D. J., and Gardner, R. C. 2000. Riding the ice age El Niño? Pacific biogeography and evolution of *Metrosideros* subg. *Metrosideros* (Myrtaceae) inferred from nuclear ribosomal DNA. Proceedings of the National Academy of Science, U.S.A. 97:4118-4123.

Wright, S. D., Yong, C. G., Wichman, S. R., Dawson, J. W., and Gardner, R. C. 2001. Stepping stones to Hawaii: a trans-equatorial dispersal pathway for *Metrosideros* (Myrtaceae) inferred from nrDNA (ITS + ETS). Journal of Biogeography 28:769-774.

Zeilinga de Boer, J., and Sanders, D. T. 2002. Volcanoes in Human History. Princeton University Press, Princeton, New Jersey, pp. 180-181.

Ziegler, A. C. 2002. Hawaiian Natural History, Ecology, and Evolution. University of Hawai'i Press, Honolulu.

Krakatau, revegetation , xix

l

Lahaina, town, 15
Lanilini Peak, Maui, 69
Lili'uokalani (Queen), xvii
Limahuli Gardens, 29, 106
Line Islands, 49
Lō'ihi, 7, 13
Lobelia relatives on the islands, 55-63

m

Macromolecular studies
    Bidens, 49
    Gossypetin, 51
    Lepidium, 55
    Lobelioids, 61-62
Metrosideros, 64
Mints, 73
    Sanicula, 75
    Scaevola, 120
    Silene, 77
    Silversword alliance, 46
    Tetramolopium, 78
    Vaccinium, 82
    Viola, 83
Mākahona Volcano, 7
Mints on the islands, 66-73

n

National Tropical Botanical Garden, 16, 29, 103
Northwestern islands (by distance)
    Nihoa, 22
    Necker Island, 22
    French Frigate Shoals, 23
    St. Rogatier & Perouse, 27
    Gardner Pinnacles, 23
    Maro Reef, 23
    Laysan Island, 23-24
    Lisianski Island, 24-25
    Pearl and Hermes Reef, 25
    Midway Atoll, 25-26
    Kure Atoll, 26

o

Observatories
    Mauna Kea Observatories   8
    Mauna Loa Observatory   10
Volcano Observatory, 3, 4, 39
Orchids, 133-136
Outrigger canoe construction, 99

p

Parks, government supported, 181
Pea family on the islands, 97-101
Penguin Bank, 19

Plant
    Communities, 9
    Dispersal, means of, xv, xviii, xxi, 65, 74, 88,
    120, 135, 168
    Names, hierarchy of, xxi
    Names, origins of, 35, 44, 101
Populations, defined, xxii
Plant reproduction, see Reproductive strategies
Polihale State Beach/Park, 21, 127
Pololū Valley, Hawai'i, 7
Puaulu, kipuka, xi, 99, 107, 118
Pu'u Huluhulu, kīpuka, xi, 10, 121, 122
Pu'u Kukui Peak, East Maui, 15

r

Reproductive strategies LISTED
Reproductive strategies, xix, xx, xxi, 110
Ring Road in the National Park, 4
Roads
    Chain of Craters Road, 4
    Ring Road, 4
    Saddle Road, 9
Rose family on the islands, 107
Rutaceae (Citrus family) on the islands, 114

s

Sailing ships
    Blonde, H.M.S., xiii, 35
    Discovery, H.M.S., xiii
    Resolution, H.M.S., xii
    Uranie, 126
Sandalwood trade, 17, 116
Sandalwood, bastard, 118
Sandwich Islands, 35
Ship Captains
    Cook, James, xii, xviii, 17
    Dubaut, J. E., 44
    La Pérouse, J. F. G. de, 5
    Vancouver, George, xiii, 5
    Wilkes, Chas., xiii
Silversword alliance on the islands, 32-47
Silverswords, origin and evolution, 45-47
Starbuck Island, 49
Sulfur Banks, 11
Sunflower family
    Age of pollen , xvi
    Bidens, 47-49
    Dandelion (Taraxacum), xiv
    Emilia, 172
    Hypochaeris, xv, 173
Silversword alliance, 32-47
    Tetramolopium, 28, 79
Surf boards, 99-100

t

Thurston Lava Tube, 11, 29

SCIENTIFIC NAMES INDEX

Pritchardia species, 22
Psidium cattleianum (strawberry guava), 70, 166
Psidium guajava (guava), 166
Psilotum complanatum, 153
Psilotum nudum, 153
Pteridium aquilinum, xiv, 123
Pueraria montana var. lobata, 160
Pyracantha angustifolia, 108
Pteris cretica, 150

r

Reynoldsia sandwicensis, 97
Rhodomyrtus tomentosa, 63
Ricinus communis (castor bean), 89
Rubus argutus (blackberry), 170
Rubus hawaiensis, 108, 122-123
Rubus macraei, 108
Rubus spectabilis, 108
Rubus, weedy species, 170

s

Saccharum officinalis (sugar cane), 132-133
Sadleria cyatheoides, 151
Sadleria squarrosa, 152
Sanicula arctopoides, 76
Sanicula sandwicensis, 76
Santalum album, 115
Santalum freycinetianum (sandalwood), 116
Santalum haleakalae, 117
Scaevola glabra, 119
Scaevola sericea, xx, 119
Scaevola taccada, 119
Schinus terebinthifolius, 176
Sicyos alba, 121
Sida fallax, 54
Silene hawaiiensis, 77
Silene struthioides, 77
Smilax melastomifolia, 123
Solandra maxima, 175
Sophora chrysophylla, 100-101
Spathodea campanulata, 168
Spathoglottis plicata, 134
Spermolepis hawaiiensis, 75
Sphenomeris chinensis, 152

Stenogyne cinerea, 69
Stenogyne haliakalae, 69
Stenogyne kamehamehae , 69
Stenogyne microphylla, 70
Stenogyne oxygona, 69
Stenogyne purpurea, 70
Stenogyne viridis, 69
Stereocaulon vulcani, 112
Sticherus owhyhensis, 143
Styphelia tameiameiae, 125
Syzygium sandwicensis, 176

t

Taraxacum officinale (dandelion), xiv
Tetramolopium humile, xv, 28, 79
Tetramolopium lepidotum, 79
Tetramolopium rockii, 79
Tetramolopium sylvae, 79
Tetraplasandra waialealae, 96
Tetraplasandra waimeae, 96
Tibouchina urvilleana (glory bush), 157
Tournefortia argentea, 133
Tropaeolum majus (nasturtium), 173-174

v

Vaccinium, 39
Vaccinium reticulatum (blueberry), 39, 40
Vaccinium calycinum , 80
Vaccinium cereum, 82
Vaccinium dentatum , 80
Viola langsdorfii, 83
Viola species, 83
Vitex rotundifolia, 21, 128

w

Waltheria indica, 129
Wikstroemia uva-ursi, 85
Wilkesia gymnoxiphium, 21, 42-44
Wilkesia hobdyi, 21

z

Zingiber officinalis, 168
Zingiber zerumbet, 168